BIOCHEMISTRY RESEARCH TRENDS

OCHRATOXINS

BIOSYNTHESIS, DETECTION AND TOXICITY

BIOCHEMISTRY RESEARCH TRENDS

Additional books in this series can be found on Nova's website under the Series tab.

Additional e-books in this series can be found on Nova's website under the e-book tab.

BIOCHEMISTRY RESEARCH TRENDS

OCHRATOXINS

BIOSYNTHESIS, DETECTION AND TOXICITY

DANIEL PORTER
EDITOR

nova publishers
New York

Library of Congress Cataloging-in-Publication Data

ISBN: 978-1-63483-693-7

Library of Congress Control Number: 2015948670

Published by Nova Science Publishers, Inc. † New York

CONTENTS

PREFACE

The mycotoxin ochratoxin A (OTA) is a secondary metabolite of certain fungi that may be present in food and feed. Consumption of contaminated food or feed may be harmful for humans and animals. This book aims at giving an overview of several aspects of OTA and the public and economic benefits of reducing OTA contamination in food. Other chapters focus on the use of plant extracts, essential oils and substances isolated from them for the control of ochratoxigenic fungi and ochratoxin production and its carcinogenic effects on animals.

Chapter 1 – The mycotoxin ochratoxin A(OTA) is a secondary metabolite of certain fungi that may be present in food and feed. Consumption of contaminated food or feed may be harmful for humans and animals. This review aims at giving an overview of several aspects of OTA: origin; producing species; detection methods; toxicity and bioassays; metabolism and further metabolites; legislation; reduction methods. Each of these items will be described and illustrated with examples. Especially the last decade, many investigations have been performed and published worldwide. In the scope of the existing shortage of food in parts of the world, safety is of utmost importance and improvements in rapid detection and in particular the avoidance of the presence of OTA may add to human and animal health as well as protect and promote trade.

Chapter 2 – Contamination of food and agricultural products of various types due to toxigenic mold is a serious problem. Fungi are present naturally and can contaminate food crops under favorable conditions of temperature, relative humidity, pH, nutrients availability and oxygen. Ochratoxin A (OTA) has been proved to have hepatotoxic, carcinogenic, immunosuppressant and teratogenic properties. It has been classified as possibly carcinogenic for humans (Group 2B) and affects the kidney in animals exposed to natural levels of contaminated food. Ochratoxin (OTA) is a mycotoxin that contaminates a wide variety of foods such as corn, barley, green coffee, wine, beer, cocoa, grapes, spices and dried fruits.

Plants cells (*Allium cepa*) are very sensitive to the mycotoxin OTA, as was observed at the highest concentration. Under these conditions OTA produced toxicity and cytogenetic injury. Human exposure to mycotoxins is mostly through plant foods. However, additional potential exposure may be through animal foods such as milk, cheese and meat from animals fed with mycotoxin contaminated foods. These can produce toxic waste and biotransformation products. However, while it may not be possible to attribute deaths or illnesses OTA does not mean they have no effect. In fact, in some regions of the world exposure to OTA could be higher and some populations more sensitive to the toxicological effects. Additionally levels of OTA in food frequently exceed regulatory limits. OTA monitoring and action to remove non-compliant food products from the market is a financial burden equally to the health authorities and food businesses. Therefore, there is potential for both public health and economic benefits of reducing OTA contamination in food.

Chapter 3 – The chlorophenolic mycotoxin ochratoxin A (OTA) is prevalent in food sources and represents a potential health hazard to various animals. OTA is a potent renal carcinogen in rodents and has been linked to the fatal human kidney disease, Balkan Endemic Nephropathy (BEN). Understanding the mechanism of OTA-mediated toxicity is vital for the assessment of the risk associated with its intake. Two theories have been proposed in support of mechanism of action of OTA toxicity, i.e., direct genotoxic (DNA binding of OTA moiety) and epigenetic or non-genotoxic (without any direct DNA binding of OTA moiety). OTA facilitates guanine-specific DNA adducts *in vitro* in orally dosed rat and pig kidneys. A low level of DNA lesions were observed in OTA treated animals by ^{32}P-post labeling technique suggesting that OTA may be activated into some other reactive metabolites which further react with DNA to form adducts. Oxidative stress also accounts for the OTA-mediated toxicity, thus favoring the epigenetic or non-genotoxic mechanism. The mechanisms of ochratoxicity include inhibition of mitochondrial respiration correlated with a depletion of ATP, inhibition of protein synthesis, enhanced lipid peroxidation and alteration of signal transduction. The combination of these basic mechanisms with an unusual long plasma half-life time (35-70 days), the metabolic conversion of OTA into still active derivatives and conjugates, both potentially reactive with cellular biomolecules including DNA could partially explain the multiple toxic effects of OTA including genotoxicity, cytotoxicity and carcinogenicity.

Chapter 4 – Ochratoxin A (OTA) is the second mycotoxin, followed by aflatoxins, of significance in several agricultural products. During the last decades, OTA has been receiving considerable attention worldwide because of its nephrotoxic, immunotoxic, teratogenic and carcinogenic effects on animals. This toxin was included in the group 2B by the International Agency for Research on Cancer (IARC) since it is a potential carcinogen for humans. *Penicillium verrucosum* is the main OTA-producing species in temperate and cold climates, and *A. ochraceus* and related species in warm and tropical climates. Due to their physiological differences, each of these species occupies a specific ecological niche. Recently, several studies reported on the presence of potential OTA-producing *Aspergillus* section *Nigri* species in food and feeds. *Aspergillus carbonarius* is considered to be the main producer of OTA, followed by species belonging to the *A. niger* aggregate. The presence of this toxin has been extensively reported worldwide in several products such as coffee, wines, beers, grapes, dried grapes, cereals and derivatives destined for humans and animals, oilseeds and derived products and, occasionally, in body fluids, plasm, meat and kidneys of several animal species. One strategy to reduce the entry of mycotoxins into the food chain is the use of chemical treatments during the storage of cereals, grains and oilseeds to reduce fungal growth and production of toxins. From a human health perspective, the use of antioxidants such as butylated-hydroxyanisole (BHA), propyl paraben (PP) and butylated-hydroxytoluene (BHT) are allowed by the US Food and Drug Administration (FDA) as antimicrobial agents. Such compounds are regarded as safe (GRAS) chemicals. The last known studies have shown that these antioxidants have a protective action in food since they could maintain its organoleptic properties. This chapter reviews data showing that treatments with food grade antioxidants significantly influence growth parameters, enzymes production and OTA production of ocratoxigenic strains. In addition, the incubation time, temperature, water activity and their interactions are also important factors involved in the development of these fungal species and in the accumulation of toxins.

Chapter 5 – Ochratoxins are mycotoxins produced principally by *Aspergillus* and *Penicillium* species, and they are found in commodities such as cereals coffee, grape and derived products of them. Ochratoxins exhibit nephrotoxic, immunosuppressive, teratogenic and carcinogenic proprieties, as well as Balkan Endemic Nephropathy have been attributed to this mycotoxin. The ochratoxin A (OTA) is the most potent of them.OTA removal from foods and feedstuffs is not feasible, so prevention management is the best method for controlling the contamination. However, in the case of the contamination occur, a variety of approaches including physical, biological andchemical, treatments to minimize the risk, decontaminate or eliminate the fungi have been attempted. A way to prevent or interfere with fungal growth and ochratoxins production is by use of natural antifungal agents. The use of natural compounds, like essential oils, with lower toxicity of the synthetic products can be a good alternative to the ochratoxin control. This chapter will focus on the use of plants extracts, essential oils and substances isolated from them for the control of ochratoxigenic fungi and ochratoxin production.

In: Ochratoxins
Editor: Daniel Porter

ISBN: 978-1-63483-693-7

Chapter 1

OCHRATOXIN: AN OVERVIEW

Eline P. Meulenberg*
ELTI Support VOF, Malden, The Netherlands

ABSTRACT

The mycotoxin ochratoxin A(OTA) is a secondary metabolite of certain fungi that may be present in food and feed. Consumption of contaminated food or feed may be harmful for humans and animals. This review aims at giving an overview of several aspects of OTA: origin; producing species; detection methods; toxicity and bioassays; metabolism and further metabolites; legislation; reduction methods. Each of these items will be described and illustrated with examples. Especially the last decade, many investigations have been performed and published worldwide. In the scope of the existing shortage of food in parts of the world, safety is of utmost importance and improvements in rapid detection and in particular the avoidance of the presence of OTA may add to human and animal health as well as protect and promote trade.

---

* Corresponding author: Eline P. Meulenberg. ELTI Support VOF, Ambachtsweg 5, 6581 AX Malden, The Netherlands. E-mail: E.Meulenberg@eltisupport.nl.

INTRODUCTION

Food and feed safety has increasingly gained attention of investigators, authorities, food/ feed producing facilities and consumers. Contamination may lead to health hazards and comprises a plethora of compounds and organisms. In particular, microorganisms and their secreted toxins pose a threat, where those of bacteria are often more toxic than those of fungi, the so-called mycotoxins. The last type of microorganisms occurs worldwide on crops and flourishes under circumstances of high humidity and temperatures. Mycotoxins are secondary metabolites that are widespread on food, feed and commodities produced from them. Although often not showing acute toxicity, they may lead to chronic diseases and conditions. The recognition of the toxicity of mycotoxins has led to the development of various detection methods and regulations.

In this overview, the main subject compounds are ochratoxins, secondary metabolites mainly produced by members of the species *Aspergillus* and *Penicillium*. The presence of fungi on crops and the persistence of the produced mycotoxins results to the occurrence of the compounds in food/feed and consequently, after ingestion, in the body of humans and animals. Having established more or less accurate detection methods, mycotoxins have been determined in various crops, bodily fluids and tissues. In the following, an overview will be given the origin and biosynthesis of ochratoxins focussed on ochratoxin A, its occurrence, toxicity, its metabolites, and of analytical methods for detection and quantification, legislation and measures to reduce toxin-producing fungi and transfer of the toxic product.

ORIGIN OF OCHRATOXINS

With regard to ochratoxins, this group of mycotoxins comprises four main members: ochratoxin A (OTA), B (OTB), C (OTC), and alfa (OTα), of which OTA is the most frequently occurring, the one with the most economical impact and the best studied. Ochratoxins are secondary metabolites, which means that they are not vital for the producing organism. However, when humans or animals get in contact with these compounds, they may be intoxicated, leading to more or less serious conditions, also called mycotoxicoses c.q. ochratoxicosis. *Aspergillus* and *Penicillium* are considered the main producers of the toxins, of which *Aspergillus* prevails in tropical regions and *Penicillium* in temperate regions. A list of subspecies is given by Bayman et al. (2002), Cabanes et al. (2010), el Khoury (2010)and Ostry et al. (2013) [1-4]. A picture of fruit infested with the fungi is given in Figure 1.

The biosynthetic pathway involves phenylalanine and sodium acetate as starting material and several enzymes as elucidated by Searcy et al. (1969), Yamazaki et al. (1971), O'Callaghan et al. (2003) and Färber and Geisen (2004) [5-8]. A schematic overview has been given by Huff W.E. [9] and further elaborated by el Khoury (2010) [7]. PCR has been used to appoint genes involved and which of these genes are absolutely required for ochratoxin biosynthesis. Similarly, deletion experiments have given further insight in the biosynthetic pathway. For example, Gallo et al. (2012) [10] describe that the deletion of a nonribosomal peptide synthase gene in *A. Carbonarius* eliminates the ability of the fungus to produce OTA. Chakrabortti and Liang (2015) [11] announced that they will give a lecture in London in September 2015, stating that they sequenced the whole genome of *Aspergillus* westerdijkiae and performed gene knock-out experiments to elucidate the regulation of secondary metabolism. The elucidation of the biosynthetic pathway enabled also the chemical synthesis of OTA, that may present a reference standard for analytical methods.

TOXICOLOGY

As already mentioned, ochratoxins are toxic for humans and animals. Of the various ochratoxins, OTA is the most toxic and, consequently, has been studied the most extensively.

Figure 1. Fruit infested with fungi.

Initially, the mouse bioassay was used for toxicity assessment, where compounds are added to feed and the rate of survival is determined.

It will be clear the this assay is very inaccurate.

An alternative bioassay used one day old larvae of the African catfish, *Clarias* gariepinus, that were or were not exposed to toxins. The proportional death of the larvae was used as a indication of toxicity. More sophisticated toxicity assays have been developed thereafter. It was recognized that OTA has nephrotoxic, hepatotoxic, carcinogenic, teratogenic, immunotoxic, it affects the brain and intestine, and it disturbs protein synthesis. In humans, symptoms include e.g., anemia, anorexia, discoloured skin, fatigue, headache, and a range of abnormal clinical parameters such as increased clotting time, leukocytes, neutrophils, eosinophils, etc.

These parameters are be included in diagnosis of intoxication. In addition, the presence of ochratoxins in blood and organs may be determined. Due to its long half-life, these compounds accumulate to a high degree. In a publication of 1999, an overview of the occurrence of OTA in human blood samples including country, incidences, and concentrations was given [12]. To assess the toxicity of ochratoxins at the cellular level, several assays have been developed and used. Kidney-derived cell lines, LLC-PK1 and OK, were used in a cytotoxicity assay wherein growth curves in the presence of increasing concentrations were used as an indicator of toxicity. It appeared that OTA was more toxic than OTB. Drawback of these assays is that they are quite sensitive to assay conditions [13]. Enzyme-induction assays were included in an *in vitro* toolbox for risk assessment of model molecules, including OTA [14].

The assays were based on reporter genes in genetically modified *E. coli* to assess acute toxicity based on altered gene expression profiles; cytochrome P450 in enterocytes (Caco-2 cells) to address effects on metabolism; the aryl hydrocarbon nuclear receptor (AhR; DR-Calux) was used to find effects on transcription processes; and the oestrogen receptor (ER; MMV-Luc cells) for the corresponding target genes. All additions were made in realistic exposure levels. It appeared that OTA had no effect on intestinal cells, nor in the P450 assay, nor in the AhR and nor in the ER assay.

In another study, effects of OTA alone or in combination with aflatoxin B1 and/or fumonisin B1 on cell viability of the cell lines caco-2, MDBK and 264.7 were tested using the MTT and neutral red assay [15]. OTA was shown to be the most toxic compound.

In a review of Reddy and Bhooda (2010) [16] the toxicity of OTA in animals and humans has been described with respect to nephrotoxicity/nephropathy, carcinogenesis, en liver toxicity. A further summary of toxicity is given by el Khoury (2010) [7], as well as in the IARC Monograph vol. 56 (1993) that includes carcinogenic data.

DETECTION METHODS

Fungi or moulds are visible on crops and they grow under certain circumstances of humidity and warmth. There have occurred many incidences after ingestion of 'contaminated' crops or fruits. These prompted the development of methods for the determination of the presence of the contaminating substances. Initially, the mouse bioassay was used, where mice are administered suspected preparations. In case the mice died, the preparations were designated as contaminated with toxic substance. However, this assay is absolutely not accurate. The toxic substance itself is not determined and several other causes may have led to the death of the animals.

Other methods were developed such as TLC, LC, HPLC, LC-MS, immunochemical assays, biosensors and assays and/or sensors comprising synthetic binding entities such a aptamers or MIPs. However, with regard to the analysis of mycotoxins, except the bioassay, in general, each assay has to be considered in conjunction with an extraction/purification step. Before a sample can be applied to the chosen method, the targeted compound has to be isolated into a solvent compatible with the assay system. Analytical methods for mycotoxins have been reviewed, in some publications in conjunction with regulations. In the following, the pretreatment and detection methods developed and used are described in more detail.

Occurrence

With regard to ochratoxins, it appears that these compounds have been subject to analysis and found in a whole range of samples: grains like maize, rice, barley, wheat, durum, oats, cereals, sorghum, as well as products for human and animal consumption (flour, foods, feeds); nuts (pistachios), herbs, spices, palm kernels, peanut kernels, chilis, paprika, cassava and fruits; tea and coffee (raw, green, roasted, instant); fluid samples like milk and milk products (infant formulae, baby food, cheese), wine, beers, beverages, serum, plasma, urine and surface water; meat and meat products (swine, poultry); and organs (liver, kidney, brain, spleen, hart). Because solid samples cannot be analyzed as such, they generally have to undergo at least a pretreatment before subjecting them to an analytical method.

Sample Pretreatment

Generally, mycotoxins c.q. OTA cannot be determined in samples directly, except in the case of water or bodily fluids. In particular, solid samples require a specific pretreatment. Such pretreatment involves grounding, dissolution, sieving, extraction, separation and optionally concentration in order to get the target compound into an appropriate solution and remove undesired matrix components and background interference. For sampling one may refer to the EU regulations. Especially when batches of food/feed are controlled, several samples should be taken from various sites as the toxins are not evenly distributed. The extraction step is dependent on the analytical method used thereafter and may involve liquid-liquid or solid phase extraction. Several extractants may be used and these should be compatible with the particular analytical method.

For example, acidic or alkaline buffers are used before TLC or HPLC of OTA. Other buffers with various composition are applied for SPE/SPME and SPE columns may also vary depending on the following method [17-20]. In addition, extraction solutions may contain organic solvents such as methanol, ethanol, acetonitril, chloroform, dichloromethane, water and combinations thereof. A dedicated pretreatment method is IAC. In IAC a specific antibody is covalently coupled to an activated solid phase, sample is passed through the column, contaminating substances are washed away and the target compound is eluted into an appropriate solution for further analysis. The principle of IAC has been described [21-23] and the application for OTA isolation/purification has been mentioned in many reports. In addition, IAC columns are commercially available and have been applied in numerous investigations. An overview of SPE and IAC methods used for mycotoxins is given in the review of Turner (2009) [24]. It should be taken into account that when developing and/or using a certain pretreatment method, validation according to national or international regulations and standards (AOAC, EU, FAO, FSA) has to be performed. Attention has to be given to recovery and reproducibility.

Analytical Methods

For mycotoxins in general an overview of analytical methods (TLC, HPLC, IA) was given by Crosby (1984) [25]. A book chapter dedicated to methods of analysis for OTA was written by Scott (2002) [26].

Herein TLC, minicolumns, LC, ELISA, spectrofluorometry, confirmation procedures and interlaboratory studies are mentioned. A general overview of methods and trends in mycotoxins analysis was given by Pittet et al. (2005) [27]. In comparison with the above book chapter, this review includes GC, GC-MS, LC-MS, as well a rapid screening methods, such as ELISA, dipsticks and lateral flow tests, solution fluorometry, NIR, MIR, MIP, CE, fluorescence polarization, and biosensors.

In the scope of regulations the evaluation of methods for mycotoxins determination is described e.g., by van Egmond [28-30]. Further reviews of mycotoxins analysis are given by Krska (2008) [31] wherein also multi-mycotoxin methods are included. Turner (2009) [24] presented a review of analytical methods for mycotoxins, in particular with regard to aflatoxins and ochratoxins. An extensive evaluation of the various pretreatment methods is given. In this review it is mentioned that OTA may efficiently be extracted and separated using C-18 SPE columns or immunoaffinity-like phases containing a hexapeptide. Also a multifiber SPME device may be used for high-throughput extraction and clean-up. Commercially, a Mycosep™ colomn specific for OTA has been developed. Using TLC for detection/analysis may be best performed on silicagel followed by fluorescent densitometry.

Further, examples of HPLC protocols, GC protocols and ELISA protocols are summarized in the respective tables.

Sheppard (2011) [32] published an update of developments in mycotoxins analysis over 2009-2010 for food and feed. In this review more attention is given to the use of antibodies both in extraction/clean-up and in detection/analysis.

Examples are the techniques of immunoaffinity chromatography, the SPR assay, the development of a lateral flow colloidal gold strip assay, SPE containing MIPs specific for OTA, and the fluorescence polarisation immunoassay. These techniques are discussed below.

Thin Layer Chromatography (TLC)

TLC is the oldest method used for mycotoxins. Herein a sample solution is applied onto a absorbent layer on a glass, metal or plastic support. The constituents of the sample are separated by using a suitable mobile phase. The detection occurs visually or, because OTA is fluorescent, by fluorometry. The sensitivity of simple TLC is not always high enough for low levels. An improvement may be 2D TLC to get rid of contaminating constituents.

Further, quantitative detection may be achieved by using densitometry. A sensitive (0.1 ng), cost-effective photosensor for detection of OTA on silica HPTLC plate was developed by Caputo et al. (2007) [33].

Liquid Chromatography (LC)

LC, HPLC, and UPHLC, optionally in combination with MS or MS-MS, is the analytical methods used most. Good separation is achieved and the sensitivity is high. Several reviews of analytical methods, including LC for mycotoxins, including OTA, have been published. In 1993, Scott [34] reported the method developments in the scope of the AOAC activities. Analysis of OTA was focussed on LC methods after C_{18} Sep-Pak in grains and processed products and pig serum at 5-50 ng/mL. In addition, a GC-MS method after methylation was mentioned for the first time, showing a detection limit of 0.1 ng/g in foods. The AOAC report of 1998 [35] deals with reference materials for OTA, the use of IAC before LC analysis for various coffee samples, and the development of immunochemical methods for cereal grains (d.l. 0.5 ng/g). Follow-ups of AOAC reports are published in 2000 [36, 37], 2001 [38] and later on. Valenta (1998) [39] reviews chromatographic methods, in particular HPLC, specifically for OTA in animal and human tissues and fluids. Details of matrix, extraction, clean-up, method, confirmation, D.L./Q.L. and recovery are summarized in a large table. Details of the table are described further in the text of the article.

Mass Spectrometry (MS)

MS is the ultimate method for identification of compounds giving information about their molecular structure and identity. It is often used for confirmation purposes. MS is never used as single method, but supplemental to separation with LC or GC as GC-MS, LC-MS, LC-MS-MS, RPLC-MS. Because OTA is not volatile, GC is not the method of choice and may be used only after derivatisation.

Scott (2002) [26] gives some examples of the application of GC-MS and RPLC-MS-MS of OTA for food samples at the sub-ppb level and for various matrices, respectively. LC-MS-MS is also mentioned in the review of Pittet (2005) [27], wherein this method is used both for analyzing OTA and as validation method. Remarked is the possible matrix effect and the sometimes low recovery. A more comprehensive and extensive review including LC-MS-MS can be found in the paper of Krska (2008) [31].

Interesting is the application for the simultaneous determination of different mycotoxins, metabolites and conjugates. Further advances and improvement of this method has led to the determination of up to 87 mycotoxins in mouldy food [40]. A brief listing of the application of multi-analyte LC-MS-MS to foods, beverages, milk, meat flour and cake is given by Sheppard (2012) [32] in an update of developments in mycotoxins analysis.

It is mentioned that the development of new certified reference material for OTA support analytical quality control.

Multi-analyte detection using LC-MS(-MS) has also been used for the determination of metabolites and/or masked mycotoxins. Plants that are infected with mycotoxins-producing fungi, try to get rid of these compounds by metabolising them. In such way, consumers of such plants also are exposed to these metabolites. In addition, consumers themselves have at their disposal several, mostly enzymatic systems, to metabolise or detoxify foreign toxic substances. Although the phenomenon has been described, there has not been much investigation into the toxicity of metabolites of mycotoxins in humans and animals until now. The term 'masked' means that such compounds are not included in the common detection methods wherein predominantly the mother compounds are searched. As some of these metabolites and masked mycotoxins may be even more toxic than their mother compounds, the importance of this sub-field is now recognized. Some mycotoxins and metabolites may also been bound to proteins and other components in cells. Generally, such bound compounds are not readily detected, because samples need a specific pretreatment to liberate them. The issue of masked and hidden forms of mycotoxins in food safety was described in a special issue by Galaverna et al. (2009) [41]. Significant amounts derived from DON, ZEA and FUM's were found in products of wheat and maize. A extensive review of masked mycotoxins was published by Berthiller et al. (2013) [42] giving an update of the state of the art. In this paper the metabolite of OTA, hydroxyl-ochratoxin, is mentioned as immunosuppressive as effective as OTA itself. Further in this paper there are described the various analytical aspects including extraction and clean-up, pretreatment such as hydrolysis, chromatographic methods and immunochemical methods. To get an idea of the increasing number of metabolites and masked mycotoxins found in food and feed up to date, the article of Streit et al. (2013) [43] states that 139 different compounds have been determined using LC-MS, listed in 3 large tables, including concentrations found, including few data for OTA, but not for its metabolites. In addition to the conventional methods, immunochemical assay for the detection of masked mycotoxins have been given in a review of Goryachema (2012) [44]. These will be discussed in the next section.

Immunochemical Methods

Immunochemical methods include all analytical methods wherein an antibody or fragment is involved as main component. The antibody recognizes its target compound, in this case a mycotoxin or in particular OTA, and binds it with high affinity. A diagram of an antibody is given in Figure 2. The extent of binding is a measure of the concentration in a sample. Because of the low levels in the immunochemical reaction, determination is performed indirectly by using a detectable tag covalently bound to either the antibody or the target compound (tracer). The theoretical background of immunochemistry has been subject of numerous books and manuals [45-48].

The production of antibodies or fragments thereof that are useful in immunochemical methods has also been subject of text books and manuals [49-52]. Antiserum or purified antibodies for use in immunochemical methods can be purchased commercially or, if not available, made in-house. In short: antibodies are protein components of the immune system produced in defence against intruding foreign substances. Because of their high binding and specificity to such substances, their potential for use in assays has been recognized in the second half of the last century. To this aim, desired antibodies are produced in laboratory animals such as rats, mice, chicken, goat, sheep, etc. They are isolated as antiserum or purified from the blood of immunized animals.

Such antibodies are called polyclonal and they consist of a collection of several types with varying degrees of affinity and/or specificity. As an alternative, monoclonal antibodies are produced using the hybridoma technique from single antibody-producing cells and these are characterized by having a defined specificity and affinity for their target compound. Generally, antibodies are produced when the target compound has a certain extent of dimensionality and molecular weight. OTA is a low-molecular compound (Figure 3) and consequently requires manipulation in order to make it immunogenics. The relevant techniques have been described in the above text books.

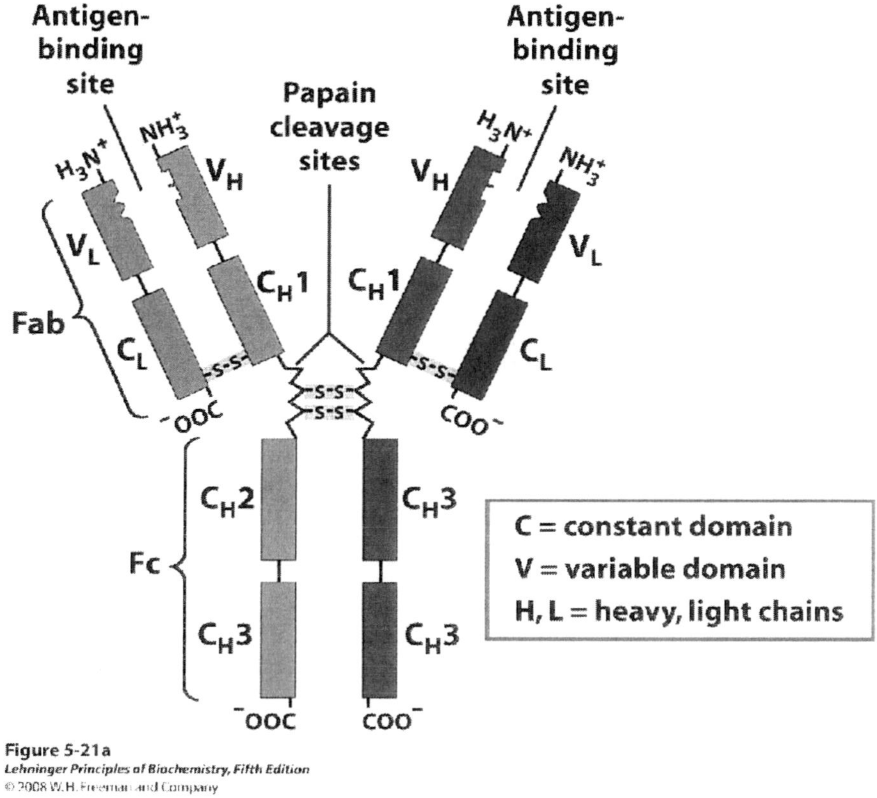

C = constant domain. V = variable domain. H, L = heavy, light chains.

Figure 2. Schematic view of an antibody.

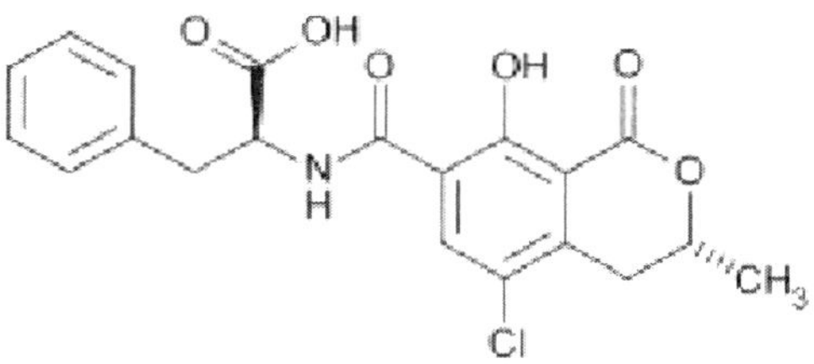

Figure 3. Molecular presentation of ochratoxin A.

The production of antibody fragments has been described in the above references and in the respective articles about immunochemical assays for the screening or detection of mycotoxins and/or OTA. A review of immunochemical methods for the analysis of mycotoxin has been published by Urusov et al. (2010) [53] and specifically for OTA by Meulenberg (2012) [54]. The first one is a short overview of immunochemical methods including antibody production and conjugates, IAC columns, enzyme immunoassays and immunosensors. The second one is an extensive overview of antibodies, immunoassay formats with various detection means, immunosensors, alternative binding entities and IAC. Antibody fragments and recombinant antibodies for the development of immunoanalytical methods for biotoxins are described by Kavanagh et al. (2015) [55]. After testing the antibody for performance, it may be used in designing a desired immunochemical assay. It is recommended to follow the European ISO norms in developing and evaluating an antibody or immunochemical assay. In this regard, especially the specificity of the antibody plays an important role for achieving a detection system for only the target compound.

Immunoassays

Most known is the immunoassay, that may be designed and named by its type of detection: radioimmunoassay (RIA), enzyme immunoassay (EIA), enzyme-linked immunosorbent assay (ELISA), fluorescent immunoassay (FIA), time-resolved fluorescent immunoassay (TR-FIA), fluorescence-polarization immunoassay (FP-IA), fluorescence resonance energy transfer (FRET) immunoassay, chemoluminescence immunoassay (CL-IS), and the enzyme-multiplied immunoassay technique (EMIT).

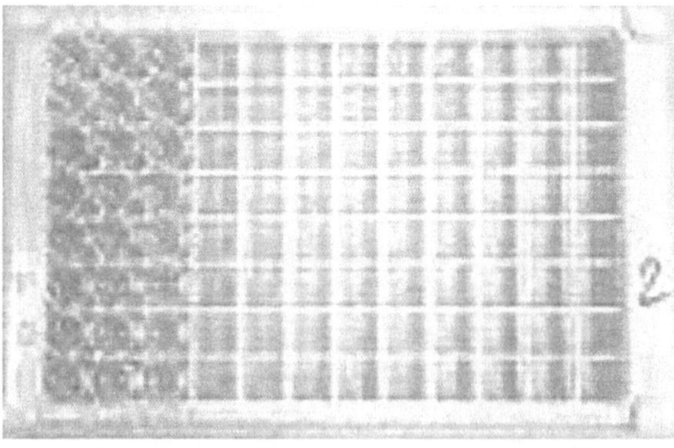

Figure 4. Image of a microtiter plate.

Most of these immunoassays are performed in microtiter plates as illustrated in Figure 4. Such immunoassays have the advantage that many samples can be measured in one run.

In addition to the above mentioned immunoassays, antibody-based lateral flow systems, dipsticks, multiplex systems and immunosensors belong to this group of analytical methods. With regard to immunoassays, there are several embodiments: reaction in solution or on a solid support. An example of the first embodiment is the RIA, wherein the tracer is the target compound synthesized to include a radioactive isotope so the both components are very similar and bind to the antibody to virtually the same extent. The most used isotopes used in RIA are ^{3}H and ^{14}C for small target compound, and ^{125}I for larger compound such as peptides or proteins. Although marking the beginning of the immunoassay technique, due to the hazards related to the handling and disposal of radioactive materials, RIA has been replaced by assays incorporating alternative tags. A ^{125}I- RIA kit has been applied for the determination of OTA in serum by Fukal and Reisnerova (1990) and Fukal (1990) [56, 57] and they found its presence in 25% of the samples at a detection limit of 100 ng/L as well as in food/food or porcine kidney. A ^{14}C-RIA was developed by Rousseau (1985, 1986) [58, 59] and applied to barley and serum at 0.5 ng/mL and 0.5 ng/mL, respectively.

The most applied immunoassay is ELISA. It is designed in either direct or indirect format, In the former, the antibody is bound to a solid support, a microtiter plate, and the immunochemical reaction is performed in the supernatant phase. The tag is an enzyme bound to the target compound that functions to generate a spectrophotometrically detectable signal used for quantification of the target compound in the sample.

In the indirect ELISA the target compound is bound to the solid phase and sample and antibody react in a competitive way. In such embodiment the enzyme tag is bound to the antibody for generation of the signal after proportional binding to the target on the solid phase. The advantages and disadvantage of ELISA are well know.

In general, samples require only minimal pretreatment as long as the target compound is dissolved in an aqueous matrix that may contain a limited amount of organic solvent. However, care should be given to samples such as plasma, serum or milk, because OTA may be bound to proteins present in the sample and may thus be overlooked.

For example, it is known that OTA is bound to BSA or HSA in plasma [60]. ELISA for OTA is included in an overview of immunochemical methods for mycotoxins [61]. Some examples are given in a table, showing format,

IC50, detection limit and MRL. A review of the application of ELISA for the determination of OTA has been published [54], summarized in a table containing formats, matrices, and detection limits. Commercially, there are many ELISA kits (qualitative and quantitative) available and these have been evaluated for performance and many other characteristics by Huybrechts and Tangni (2009) in cooperation with the kits manufacturers [62]. The comparative results are given in tables in figures.

Inherently, ELISA is performed in a laboratory environment. However, in some cases the presence of OTA in the field may be preferred. For such situations a rapid field-test may be used, e.g., a lateral flow test, a dipstick, or an IAC-assay. Lateral flow tests are based on the same principle as TLC, but containing specific antibody to capture the target at a designated site of the support, where an enzymatic reaction may generate a detectable spot. For reviews of this technique, see Bazin et al. (2010) and Posthuma 2012 [63, 64]. Dipsticks are comparable, but here the dipstick bound with antibody is immersed into the sample solution for the immunochemical reaction on the support [61, 65] (Prieto-Campas 2009). Positive samples may then be transferred to the lab for further analysis. Due to the low detection limit of such devices, they are very useful for a quick screening in the field. An evaluation of commercially available strip tests in conjunction with EC norms, is given in the review of Bazin et al. (2010) [63]. Although such rapid tests are mostly qualitative, a test strip for quantitative detection of OTA was developed by Anfossi et al. (2011) [66]. The whole system was rapid and simple. Cereal extracts in PBS + 2%PEG were applied on the strip containing gold-labelled antibody, incubated 8 min. and measured using a scanner that could provide quantitative results at a detection limit of 1.5 μg/kg. A variant immunochemical assay for both lab and field application is the IAC-assay. Herein an IAC-column as described above combines the step of extraction/isolation onto the columns with a detection step. The column is composed of 3 layers: clean-up layer, detection layer containing OTA-specific antibody, and conjugate (tracer) layer. If OTA is present in a sample, it is bound to the antibody and detected by flowing the tracer through the detection layer. The result is determined visually. This format of IAC is useful as semi-quantitative field test. However, in lab environment it has been tested for herbs and spices [67] that appeared to contain OTA above the control level of 10 μg/kg.

Membrane Immunoassay

Comparable to the IAC assay, but in a different flow-through format, a rapid immunochemical test comprises the membrane assay. Herein a suitable membrane is coated with secondary antibody and the immunochemical reactants – anti-OTA antibody/sample/ tracer/substrate - are passed through to membrane. A quantitative field test is possible when using a portable colorimeter [68]. Even difficult samples such as roasted coffee have been analysed for OTA with membrane immunoassay at a sensitivity of 4 μg/kg. The principle and application of the membrane assay is illustrated in a report about the rapid analysis of OTA in red wine [69]. In practice, membrane immunoassays have not found broad application.

Fluorescent Assays

Immunochemical assays using a fluorescent tag instead of an enzyme include FIA, TR-FIA, FP-IA and FRET. In contrast to enzyme immunoassay where the number of useful enzymes is limited, there is a large number of fluorophores that may be used as a tag in immunoassay. FIA is the most simple fluorescent immunoassay performing similar to ELISA but with fluorescence detection. With regard to OTA, no reports of this type of assay has been found. Whereas in common immunoassay a separation step is needed before detection, the change in and/or measurement of fluorescence, TR-FIA, FP-IA and FRET omits the separation step contributing to speed and simplicity of the assay. TR-FIA uses a fluorophore with time-resolved detection giving the advantage that background fluorescent is largely avoided. Its application for OTA has been described by Huang et al. (2009) [70]. The design included coating of OTA-BSA onto microtiter plate, adding specific antibody, followed by secondary antibody labelled with Sm^{3+}. Addition of a fluorescent intensifier further enhanced the sensitivity to 0.05 μg/L. FP-FIA as a rapid screening assay for mycotoxins was presented at workshop and described by Maragos (2006) [71], followed by a review of this author [72]. Its application to the detection of OTA in a screening of food samples after simple methanol extraction is found in an article of Shim et al. (2004) [73]. A detection limit of 3 ng/mL was achieved. Red wine as matrix for FP-IA was the subject of a study by Zezza et al. (2009) [74] for high-throughput screening (results in less than 10 minutes) showing a detection limit of 0.7 ng/mL of OTA. FRET is a label-free method wherein the intrinsic fluorescence of antigen and antibody complex is analyzed in solution. A FRET immunoassay for the quantitative detection of OTA was developed by Taihua et al. (2011) [75]. The technique is illustrated for the detection of aflatoxin B1, showing that Fab antibody fragment provided better limit of detection than whole antibody [76].

Multi-Analyte Immunoassay (MAI)

In comparing the various detection methods, a disadvantage often mentioned for immunoassays is the feature that only one target compounds is determined, or similar substances without discrimination, whereas in LC methods various compounds may be determined in one run. However, an advancement was the development of multi-analyse immunoassays, MAI. This embodiment is characterized by the ability to detect several compounds in the same assay. Such an assay is especially suitable to determine the co-occurrence of different mycotoxins in the same sample or to include metabolites and masked metabolites of OTA. In principle, the ELISA is not suitable for MAI as also stated in the review of Goryacheva et al. (2007) [77]. It may be feasible when e.g., different rows of a microtiter plate are coated with antibodies specific for different compounds and a mix of the corresponding tracers are added. In our own lab we developed a MAI for four organophosphate pesticides that could be measured simultaneously. No such embodiment has been found for mycotoxins and/or ochratoxins in literature. As the choice of fluorophores with different fluorescence is much larger, the development of MAI-FIA may be more promising. Huang et al. (2009) [70] presented a dual-label TR-FIA wherein both aflatoxin B1 and OTA are detected simultaneously using Eu^{3+} and Sm^{3+} coupled to secondary antibody, respectively. An alternative MAI embodiment for simultaneous detection of aflatoxin B1 and OTA on IAC column has been described [78]. By combining one clean-up layer with two detection immunolayers containing secondary antibody, and anti-aflatoxin B1 and anti-OTA in each detection layer, a tandem immunoassay column was constructed. Such columns could be used for rapid screening of spice samples at cut-off levels of 5 and 10 μg/kg, respectively. A similar MAI column, but for four mycotoxins (OTA, FumB1, DON, ZEA) has also been reported.

In this gel-based format the different monoclonal antibodies were each covalently coupled to activated Sepharose 4B and put onto each other divided by a frit. Sample mycotoxins were flowed through the column, followed by a mixture of HRP-conjugates of the respective mycotoxins, and then substrate. This semi-quantitative affinity column may be used for screening of contaminated samples showing cut-off values of 3, 1,250, 1,000 and 200 μg/kg for each of the above mycotoxins, respectively. These investigators used the same immunoreagents to produce a membrane-based flow-through MAI incorporated in a plastic cassette and onto an absorbent pad. Reading of the results was visually and showed cut-off values of 3, 5, 700 and 175 μg/kg, respectively.

A rather recent development of MAI is the multiplex method using the Luminex xMAP system (Luminex Corp., Austin, TX, US). This assay system uses sets of microbeads incorporating two fluorescent dyes that serve as a coding means for identification of the beads. In a study of Anderson et al. (2010) [79] this system was used for the detection of two mycotoxins, FumB1 and OTA, in grain products. The beads were coated with target molecule by covalent binding. In the wells of a microplate the target compounds were added, followed by the bead set. Then biotinylated monoclonal antibodies were added as well as streptavidin-phycoerythrin (signal-generating tag). The mixtures were analyzed using the Luminex 100 flow cytometer and corresponding software. In this study OTA and FumB1 could be detected simultaneously in spiked oats and cornmeal at low levels. A comparable multiplex assay on the Luminex system, but including six mycotoxins (Afl, OTA, DON, Fum, ZEA, T-2 toxin) as target compound for detection in feed was reported by Peters et al. (2011) [80]. A similar assay for wheat and cereals was described by Aqai et al. (2011) [81]. The microbeads used were superparamagnetic MagPlex® and the different steps of the assay were performed after magnetic separation. Mycotoxin-BSA conjugates were covalently coupled to the beads, added to the wells of a microplate and incubated with sample plus specific antibody, followed by R-PE conjugated secondary antibody. Measurement was performed in a Luminex flow cytometer and results were evaluated by the StarStation software and confirmed with LC-MS/ MS. Standard curves provided IC_{50} values of 0.29, 0.33, 0.39, 1.6, 2.2 and 6,7 ng/mL for OTA, AflB1, ZEA, FumB1, T-2 and DON, respectively.

Further evaluation using spiked rape seed meal showed that this multiplex assay is suitable for analysis according to EU guidance levels. A further flow cytometry based immunoassay based on competitive fluorescent microsphere immunoassay for multi-analyte analysis of mycotoxins in grains, including OTA, was described by Czeh et al. (2012) [82]. The detection limits reportedly exceeded those of standard ELISA. For OTA a detection limit of 1.50 μg/kg was found, against maximum EEC levels of 5 μg/kg.

Synthetic Binding Entities

The antibodies that are used as core component in immunochemical methods, are proteins and consequently should be checked for performance and deterioration. Further, the reproducibility of production may pose problems, whether polyclonal or monoclonal. Alternative haven sought and two different solutions have been found: MIP and aptamers. A MIP for zearalenone (ZEA) showing high affinity and broad selectivity for ZEA and related compounds has been synthesized and used for clean-up and preconcentration of such toxins in food samples (Urraca et al. 2006) [83]. MIPs for OTA are subject of a review of Yu and Lai (2010) [84]. An overview is given of the different monomers used in the course of time, preparation strategies, its application in solid phase extraction for single analytes and group-selective sorbents as well in LC-MS/MS analysis of OTA and metabolites. In conclusion it was stated that MIPs may be regarded as synthetic receptors with high-affinity and selectivity. They are thermally stable, chemically inert and show a long shelf life at room temperatures and humidity. However, due to high costs in large-scale preparation, antibodies still are preferred.

Aptamers are a completely different class of compounds. They consist of DNA or RNA of limited length and are characterised by an affinity for target compounds equal to or higher than those of antibodies. Aptamers are produced following a repeated process of selection and amplification. It starts with a large library (10^{10}-10^{16}) of nucleic acids fragment of 20-80 nucleotides, flanked by primer sequences. The library is added to the target compound preferably bound to solid phase or magnetic particles. After incubation and washing the bound aptamers are eluted and subjected to further rounds under conditions of increasing stringency. This whole process is called SELEX (Systematic Evolution of Ligands by Exponential Enrichment) and finally results in sequences with high affinity to the target compound that may be chemically produced in large quantity once their sequence has been elucidated. Recent reviews of the generation of aptamers and their biological application have been given by Mairal et al. (2008), Zhou et al. (2010) and Song et al. (2012) [85-87]. As with antibodies aptamers may be applied in various methods: in IAC for specific isolation and concentration; in analytical methods comparable to immunoassays; in diagnosis; in sensors; and even as therapeutics. A review of aptamers specifically for OTA has been published by Rhouati et al. (2013) [88]. In this report, for example, the properties of aptamers and antibodies are compared and aptamers are considered good alternatives in bioanalytical methods. SELEX was used for the identification of a DNA aptamer for the determination of OTA in naturally contaminated wheat samples [89]. The aptamer with the highest affinity was bound to an affinity column to isolate OTA from sample extracts and the concentration was determined based on the fluorescence in the eluate. There was no cross-reactivity (< 1%) with OTB. An assay named ELAA (ELISA-like enzyme-linked aptamer assay) in direct and indirect format for the detection of OTA in spiked red wine was developed, evaluated and compared with an MAb-based ELISA. The performance of the ELAA was found good and showed a detection limit of 1 ng/mL, which makes it a useful screening method (Barthelmebs et al. 2011) [90]. Whereas the above assay takes about 125 minutes, an comparable but shorter assay in microtiter plate was developed by Yang et al. (2013) [91]. Herein three oligonucleotides are involved: an OTA-aptamer, a DNAzyme, and a blocker. At binding of OTA to its aptamer, the DNAzyme is unblocked and an HRP-like enzyme activity opens up. TMB is then used for colour development. In such assay a detection limit of 4 nM in wine samples was achieved. A further application of an aptamer for OTA assay is described by McKeague and DeRosa (2012) [92]. The aptamer consists of an OTA-binding part and a double-stranded part, the latter being used for the intercalation of SYBR Green I. Upon binding of OTA, SYBR is released and determined by fluorescence detection. This label-free fluorescent sensing system showed a limit of detection of 9 nM and was highly specific. If performed in microtiter plate, many samples may be measured in a short time. An OTA-aptamer based fluorescent strip test for rapid screening of samples for contamination was reported by Wang et al. (2011) [93]. The configuration and principle of this strip test was comparable to the antibody-based strip tests described above. The competitive assay involved QD-labelled aptamer that was run on the membrane strip together with sample, passing a test zone and control zone with different DNA probes for hybridisation.

Visual detection provided qualitative results, but using particular software enabled also quantitative evaluation with a limit of detection of 1.9 ng/mL. In addition to the above embodiments, aptamers for OTA have also been applied in biosensors. These will be discussed in the next section.

Biosensors

Sensors are devices that ‘sense’ the presence of anything the user is interested in. Biosensors are defined as ‘...an analytical device incorporating a biological material, a biologically derived material, or a biomimic, intimately associated with or integrated within a physicochemical transducer or transducing microsystem...’ The ideal biosensor provides real-time results and optionally is a portable field system. The biological or biomimic part includes enzymes, receptors, nucleic acids, antibodies or fragments, microorganisms, MIPs or aptamers. Transduction systems include electrochemical (potentiometric, amperometric, impedimetric) systems, piezoelectric, gravimetric or microcantilever systems, SPR, acoustic or capacitance systems. The principles and background of the systems have been explained in several reviews [94-99]. At the sensing side of the device resides the biomolecule recognizing the target compound. Essential in the design of biosensors is the choice of biomolecule - such as a protein, receptor, enzyme, DNA, aptamer, antibody, antibody fragment, MIP - the choice of surface material, and the method of immobilization of that biomolecule such that the immobilization method does not influence the biological activity, whereas the recognizing part is free for reaction. A comprehensive overview of the aspects of immobilization of biomolecules for sensor development is given by Albers et al. (2001) [100]. In the scope of the present review, only biosensors developed for mycotoxins and in particular OTA will be discussed. It appears that two types of biosensor for OTA have been developed: antibody-based biosensors or immunosensors and aptasensors. An immuosensor for the fluorescent detection of OTA in barley extracts is mentioned in the review of Goryacheva (2009). The study of Ngundi et al. (2006) [101] referred to in this article, describes the development of a flow cell array consisting of waveguide slides coated with NeutrAvidin and bound with biotine-OTA. Passing sample and Cy5-anti-OTA resulted in fluorescent signals correlated with the concentration of OTA using an array biosensor. Testing food samples revealed LODs between 1 – 180 ng/g depending of the type of sample. Disavantage of this biosensor is that reproducible reuse is not feasible. An SPR immunosensor using nano-size gold hollow ball (GHB) and monoclonal anti-OTA antibody has been described by Fu (2007) [102]. Here the antibody is coated onto the chip surface by means of GHB conjugation. Binding of OTA resulted in a clear shift in the resonant angle. The concentration range for OTA detection was 0.05-7,5 ng/mL with a detection limit of 0.01 ng/mL. As real samples spiked milk was tested. This biosensor appeared to show good precision, the assay was simple and rapid and the chip was reusable. A different type of biosensor involved electrochemical detection (Liu et al. 2009) [103]. Herein a gold electrode with a gold colloid layer served for loading of OTA-ovalbumin. The immunochemical reaction was carried out by adding sample + monoclonal anti-OTA, followed by ALP-labelled anti-mouse IgG and substrate.

Read-out of the signal was performed using an electrochemical workstation. The OTA concentration range was 10 pg/mL-100 ng/mL, with a detection limit of 8.2 pg/mL. Corn samples showed negligible matrix effects and good recovery.

The assay took about 1 hour to complete. A highly comparable biosensor assay was developed by Yuan et al. (2009) [104] for measurement on the Biacore (Sweden). This sensing system is based on surface plasmon resonance (SPR) as illustrated in Figure 5.

On gold nanoparticles with SAM surface was immobilized a new conjugate, OTA-PEG-OVA. Monoclonal anti-OTA antibodies were used for the immunochemical reaction. Optimization of the biosensor resulted in LODs for OTA in oats and corn of 0.3 and 0.5 ng/g, respectively. Remarkable of the biosensor is that the chip can be used in over 600 binding/ regeneration cycli. While the biosensors described above use the indirect assay, a different biosensor for OTA with electrochemical detection in the direct format was described by Radi et al. (2009) [105]. A gold electrode containing a layer of 4-nitrophenyl enabled the covalent binding of polyclonal anti-OTA antibodies. Competition between sample and OTA-HRP was followed by the addition of TMB for the generation of the electrochemical signal with a detection limit of 12 ng/mL. No real samples were tested. Zamfir et al. (2011) [106] developed immunosensors based on magnetic nanoparticles coated with monoclonal antibodies and various gold electrodes. Several types of signal processing (cyclic voltammetry, electrochemical impedance spectroscopy, SPR) were tested. The SPR sensor for OTA was compared to the EIS biosensor and ELISA for the detection of OTA in white wine samples and both biosensors showed high sensitivity, precision and reproducibility, and thus are promising bio-analytical tools for food quality control.

A multiplex microarray SPR assay for the detection of multiple mycotoxins is also feasible based on the results of a study of Dorokhin et al. (2011) [107]. The principle was demonstrated based on the simultaneous detection of DON and ZEN. No further application of this technique for OTA has been found. An SPR biosensor for multiple mycotoxin analysis including OTA, which was a real biosensor in that the analysis time was short and the sensor has the potential to be designed as a field device based on the Spreeta (Texas Instruments), was described by van der Gaag et al. (2003) [108]. With regard to OTA a detection limit of 0.1 ng/g was achieved, which is well within the advisory levels.

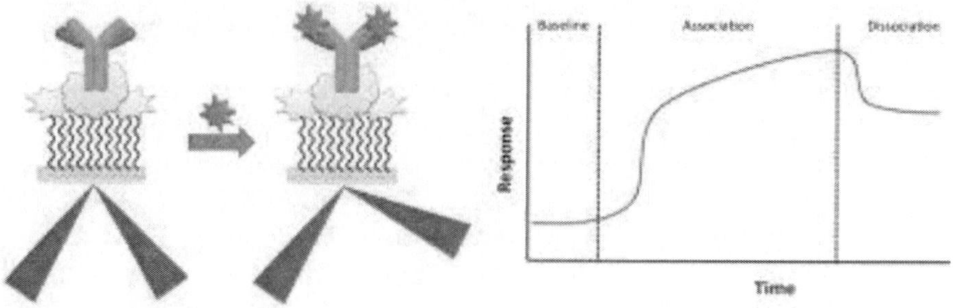

Figure 5. Schematic presentation of the principle of SPR.

Aptasensors are a relatively new development in biosensor technology characterized by the use of specific aptamers as capturing or signal-generating entities.

Palchetti and Mascini (2012) [109] published a review of electrochemical nanomaterial-based aptasensors, describing methods to integrate aptamers in biosensor applications using different nanomaterials and nanoparticles. Assays may be divided in direct, competitive or sandwich format comparable to immunoassays and depending on the size of the target molecule. Different metallic nanoparticles quantum dots (QD), graphene and carbon carbon nanotubes and magnetic nanoparticles, and their features and applications are discussed.

Electrochemical aptasensors specifically for OTA are also described. In 2010, Kuang et al. Reported a "signal-off" sensor for OTA comprising 3 DNA molecules: a linker DNA bound to nanoparticles, aptamer containing a segment complementary to the linker DNA, and AU NP-functionalized DNA complementary to a segment of the aptamer. Methylene blue, that binds to all three DNAs, was used as the sensing probe. Binding of OTA to the aptamer released the aptamer and the AU-DNA into solution, and changed the peak current due to MB. Several assay conditions were optimized to provide rapid and sensitive (30 pg/mL) detection of OTA, and the sensor was applied to spiked red wine samples to confirm applicability and accuracy in comparison with ELISA. A variant EC aptasensor is described by Tong et al. (2011) [110]. In their embodiment a gold electrode was bounded with ferrocene labelled ssDNA and hybridized with OTA aptamer. Binding of OTA resulted in release of aptamer and hairpin formation of the ssDNA, which could be followed with a change in peak current. Using an exonuclease enabled reuse of the sensor. Although a detection limit of 1.0 pg/mL could be achieved and the aptasensor performed well with wheat starch samples, the procedure itself is long.

Another quite simple embodiment of an electrochemical aptasensor is described by Bonel et al. (2011) [111]. Herein magnetic beads were coated with streptavidin and then biotine-conjugated OTA-aptamer was immobilized. On this surface a competition reaction between OTA and OTA-HRP was performed.

Addition of substrate the enzymatic product was determined using differential-pulse voltammetry, showing a detection limit of 0.07 ng/mL. Performance was tested with certified and spiked wheat samples, showing that this biosensor provides a very rapid and useful assay.

REGULATIONS

The number of countries with regulations for mycotoxins in food and feed is still increasing. Regulations are the results of a cascade of events and actions. It starts with the occurrence of incidences of intoxication of toxins, followed by the assessment of the causative agent/compound(s). Then investigations are extended, including the development of accurate analytical methods, determination of the origin of the particular compounds, its presence in/on crops, commodities, consumable parts of animals, body fluids and organs. In parallel, toxicity studies are performed at several levels. Also surveys are performed in order to determine the extent of contamination and the presence in animals and humans. Analytical methods have to be accepted, implemented and harmonized between countries [112].

Based on the results of the above steps, regulations are established nationally and internationally, in this case for mycotoxin-food/feed combinations, including maximum levels. The aim of such regulations is to protect humans and animals against the exposure to harmful substances, to prevent losses and to promote trade. Regulations are important for authorities and legislators, producers and trade organisations. Further surveys are designed to control food/feed/commodities for compliance with the existing regulations. Organisations involved include the CEN, AOAC, HACCP, and for Europe are found in the respective EU directives as well as regulations in non-EU countries. An overview of EU regulations has been given by van Egmond et al. (2007) [30]; for Morocco see the publication of Zinedine and Manes (2009) [113]. Among the different mycotoxins, for aflatoxins EU regulations exits, whereas for DON, ZEA, FUM and OTA recommendations have been established [43]. This latter publication is a review of the state of the art with regard to the current situation of mycotoxin contamination and co-occurrence in animal feed. It contains several tables illustrating the results of surveys published and studies of mycotoxin co-occurrence in food and feed ingredients in European countries. It appears that OTA is present in combination with other mycotoxins in feed and raw materials in many European countries.

As mentioned above, a rather new field of mycotoxin investigations is the determination/ analysis of masked mycotoxins. As these masked toxins are found in food and feed, and thus are ingested by humans and animals, they may pose an hitherto unknown problem. Although generally less toxic then the parent compound, they may be converted back in the body and then as yet exert their toxic activity. Berthiller (2013) suggested to subject masked mycotoxins to new regulations.

Control/Reduction

As is clear from the above, many analytical methods have been developed to measure and quantify mycotoxins in food, feed and all kinds of commodities. One of the purposes consisted of methods with a working range and detection limit complying with the regulations set for the particular compounds. However, these methods can be applied when the toxins are already present. On the other side, because mycotoxins and in particular OTA is highly stable and persistent, measures to control and reduce them have also gained much attention. These comprise especially practices during the growth of crops and in storage. As mycotoxigenic fungi prefer certain conditions of humidity and heat, the first thing to take into account is the period of sowing, planting and harvesting. Alternating crops may also lead to a reduction of harmful fungi. Using fungicides might reduce their growth, but is always accompanied by side effects. Further, the addition of other microorganisms and fungi that compete for nutrients, may lower the number of harmful fungi. Before harvest there may be performed also the sorting to remove damaged and clearly affected products. After harvest, correct storage of crops may prevent further growth. Storage should be under dry circumstances with low oxygen levels, if possible. For example, the FSA has published the UK code of good storage practices to reduce OTA in cereals. In the next step, the processing of harvested products, especially for the production of feed and beverages, numerous attempts have been made and measures have been tested to lower the level of mycotoxins or convert them to less toxic metabolites. In the article of Binder (2007) [115] is mentioned the addition of absorbents to bind the mycotoxins, so-called mycotoxin binders. Further, enzymatic and microbial detoxification are useful to counteract the impact of fungal toxins. Beverages that are consumed world-wide in large quantities are beer and wine, and from large surveys it appears that OTA may be present in most of them. In addition to pesticides, biocontrol includes the use of epiphytic yeast, sorting out moldy grapes before wine making, the addition of binders and activated carbon during the wine making as well a certain bacteria and inactivated yeast cells [116]. The effect of several binders on the analytical performance in the animal feed has been investigated and found to have only a low influence [117]. With regard to OTA, an overview of several prevention and protection methods is described [118]. The possible measures for grains are: rapid drying and store at a dry place under controlled conditions, the use of food additives as, for example, antioxidants to reduce damage; removing or reducing OTA with physical, chemical and biological methods; physical methods using absorbent materials; chemical methods including different salts, although these substances should be use with care because of the possibility of the production of unknown products and the influence on taste and quality; microbiological methods for the decomposition, absorption and conversion of OTA.

The control and prevention of the presence of mycotoxins in food and feed as well as means of protection against adverse effect, including numerous methods and substances have been extensively reviewed [117, 119]. The interested reader is referred to these publications. The former describes in detail the different methods (practical measures, physical, chemical and biological) and substances (pesticides, organic and inorganic mycotoxin binders, biological substances such a bacterial and yeast components, enzymes, anti-oxidants, etc.) for decontamination and detoxification. The latter is focussed on the inhibition of biological activities of selected toxins that may be present in the diet by natural compounds. With regard to OTA a small section is included that mentioned the application of anti-oxidants, antimicrobials, bacteria and yeasts. However, it should be remarked that even when applying such measures, surveys and analysis shows that products such as e.g., spices and nuts, still contain mycotoxins including OTA [120, 121].

CONCLUSION

Since centuries it is known that our daily food may lead to diseases and many protection measures have been taken for preservation. The recognition that certain fungi may infect crops as well as stored food/feed and that some of their excreted compounds can have serious harmful effects on animals and humans, has been a stimulus for numerous investigations and the establishment of national and international regulations. The occurrence of incidents such as the Balkan endemic has revealed that mycotoxins are the causative substances.

One of these, ochratoxin A, is the subject of this chapter. Herein ochratoxin is decribed from its origin to its toxicity to detection methods, regulations and control. The objective of this chapter is to give a state of the art regarding this toxic compound. It will however be clear that in view of the enormous literature concerning the subject, this overview has not been described exhaustively, but the reader may get an idea about what is going on. Probably, contamination with mycotoxins and in particular OTA will never be completely eliminated, but further research and better control may limit its presence to an extent that is acceptable. Future research should be focussed on further elaboration of regulations and harmonization between countries as well as rapid and cheap detection methods available for third countries.

REFERENCES

[1] Bayman, P., Baker, J. L., Doster, M. A., et al. Ochratoxin production by the *Aspergillus ochraceus* group and *Aspergillus alliaceus*. *Appl. Environ. Microbiol.* 2002, 68: 2326-2329.

[2] Cabanes, F. J., Bragulat, M. R., Castella, G., Ochratoxin A producing species in the genus *Penicillium*. *Toxins* 2010, 2: 1111-1120.

[3] Khoury el, A., Atoui, A., Ochratoxin, A., A genera loverview and actual molecular status. *Toxins* 2010, 2: 461-493.

[4] Ostry, V., Malir, F., Ruprich, J., Producers and important dietary sources of ochratoxin A and citrinin. *Toxins* 2013, 5: 1574-1586.

[5] Searcy, J. W., Davis, N. D., Diener, U. L., Biosynthesis of ochratoxin A. *Microbiol.* 1969, 18: 622-627.

[6] Yamazaki, M., Maebayashi, Y., Miyaki, K., Biosynthesis of ochratoxin A. *Tetrahedron Lett.* 1971, 25: 2301-2303.

[7] O'Callaghan, J., Caddich, M. X., Dosbon, A. D. W., A polyketide synthase gene required for ochratoxin A biosynthesis in *Aspergillus ochraceus*. *Microbiol.* 2003, 149: 3485-3491.

[8] Färber, P., Giesen, R., Analysis of differentially-expressed ochratoxin A biosynthesis genes of *Penicillium nordicum*. *Eur. J. Plant Pathol.* 2004, 110: 661-669.

[9] Huff, W. E., Hamilton, P. B., Mycotoxins – Their biosynthesis in fungi: Ochratoxins – Metabolites of combined pathways. *J. Food Prot.* 1979, 10: 778-835.

[10] Gallo, A., Bruno, K. S., Solfrizzo, M., et al. New insight into the ochratoxin A biosynthetic pathway through deletion of a nonribosomal peptide synthase gene in *Aspergillus carbonarius. Appl. Environ. Microbiol.* 2012, 78: 8208-8218.

[11] Chakrabortii, A., Liang, Z-H., Biosynthesis of ochratoxin A in *Aspergillus westerdijkiae.* 2015, *17th Conference on Food Process Engineering*, ICFPE London.

[12] Peraica, M., Radic, B., Lucic, A., et al. Toxic effects of mycotoxins in humans. *Bull. WHO* 1999, 77: 754-766.

[13] Bondy, G. S., Armstrong, C. L., Cytotoxicity of nephrotoxic fungal toxins to kidney-derived LLC-PK_1 and OK cell Lines. *Cell Biol. Toxicol.* 1998, 14: 323-332.

[14] Ribonnet, L., van der Heiden, E., Nobels, I., et al. Potential of an *in vitro* toolbox combined with exposure data as a first step for the risk assessment of dietary chemical contaminants. *Food Addit. Contam.* 2011, 28: 1136-1158.

[15] Clarke, R., Connolly, L., Frizzell, C., Elliott, C. T., Cytotoxic assessment of the regulated, co-existing mycotoxins aflatoxin B1, fumonisin B1 and ochratoxin, in single, binary and tertiary mixtures. *Toxicon* 2014, 90: 1-12.

[16] Reddy, L., Bhoola, K., Ochratoxins – Food contaminants: Impact on human health. *Toxins* 2010, 2: 771-779.

[17] Terada, H., Tsubouchi, H., Yamamoto, K., et al. Liquid chromatographic determination of ochratoxin A in coffee beans and coffee products. *J. Assoc. Off. Anal. Chem.* 1986, 69: 960-964.

[18] Akyama, H., Chen, D., Miyahara, D., et al. A rapid analysis of ochratoxin A in coffee beans and cereals. *J. Food Hyg. Soc. Jap.* 1997, 38: 406-411.

[19] Sibanda, L., De Saeger, S., Barna-Vetro, I., Van Peteghem, C., Development of a solid-phase cleanup and portable rapid flow-through enzyme immunoassay for the detection of ochratoxin A in roasted coffee. *J. Agric. Food Chem.* 2002, 50:6964-6967.

[20] Monaci, L., Pamisano, F., Determination of ochratoxin A in Foods: State-of-the-art and analytical challenges. *Anal. Bioanal. Chem.* 2004, 378:96-103.

[21] Hage, D. S., Affinity chromatography: A review of clinical applications. *Clin. Chem.* 1999, 45: 593-615.

[22] Stevensen, D., Immuno-affinity solid-phase extraction. *J. Chromat. B: Biomed. Sci. Appl.* 2000 745: 39-48.

[23] Bhargav, K., Vineeta, K., Vilasrao, K., Immunoaffinity chromatography. *Int. J. Res. Pharm. Chem.* 2012, 2: 346-359.

[24] Turner, N. W., Subrahmanyam, S., Piletsky, S. A., Analytical methods for determination of mycotoxins: A review. *Anal. Chim. Acta* 2009, 632: 168-180.

[25] Crosby, N. T., Review of current and future analytical methods for the determination of mycotoxins. *Food Addit. Contam.* 1984, 1:39-44.

[26] Scott, P. M., Methods of analysis for ochratoxin A. In: *Mycotoxins and food safety*. Ed. J. W. de Vries, M. W. Trucksess, L. S. Jackson. 2002 Plenum Publishers, New York/ Boston/Dordrecht/Moscow. Pag. 117-134.

[27] Pittet, A., Modern methods and trends in mycotoxin analysis. *Mitt. Lebensm. Hyg*. 2005, 96: 424-444.

[28] Egmond van, H. P., Methods for determining ochratoxin A and other nephrotoxic mycotoxins. *IARC Sci. Publ.* 1991, 115: 57-70.

[29] Egmond van, H. P., Analytical methodology and regulations for ochratoxin A. *Food Addit. Contam.* 1996, 13 Suppl.: 11-13.

[30] Egmond van, H. P., Regulations relating to mycotoxins in food. *Anal. Bioanal. Chem.* 2007, 389: 147-157.

[31] Krska, R., Schubert-Ullrich, P., Molinelli, A., et al. Mycotoxin analysis: An update. *Food Addit. Contam.* 2008, 25: 152-163.

[32] Sheppard, G. S., Berthiller, F., Burdaspal, P., et al. Developments in mycotoxin analysis: an update for 2009-2010. *World Mycotoxin J.* 2011, 4: 3-28.

[33] Caputo, D., de Cesare, G., Fanelli, C., et al. Innovative detection system of ochratoxin A by thin film photodiodes. *Sensors* 2007, 7: 1317-1322.

[34] Scott, P. M., Mycotoxins. *J. AOAC* 1993, 76: 112-119.

[35] Trucksess, M. W., Mycotoxins (Committee on natural toxins). *J. AOAC Int.* 1998, 81: 128-137.

[36] Trucksess, M. W., Mycotoxins (Committee on natural toxins). *J. AOAC Int.* 2000, 83: 442-448.

[37] Trucksess, M. W., Maragos, C. M., Joint Mycotoxin Committee (Technical Committee Reports). *J. AOAC Int.* 2000, 83: 536-541.

[38] Trucksess, M. W. Mycotoxins. *J. AOAC Int.* 2001, 84: 202-211.

[39] Valenta, H., Chromatographic methods for the determination of ochratoxin A in animal and human tissues en fluids. *J. Chromat. A* 1998, 815: 75-92.

[40] Sulyok, M., A liquid chromatography/tandem mass spectrometric multi-mycotoxin method for the quantification of 87 analytes and its application to semi-quantitative screening of moldy food samples. *Anal. Bioanal. Chem.* 2007, 389: 1505-1523.

[41] Galaverna, G., Dall'Asta, C., Mangia, M., et al. Masked mycotoxins: an emerging issue for food safety. *Czech J. Food Sci.* 2009, 27: S89-S92.

[42] Berthiller, F., Crews, C., Dall'Asta, C., et al. Masked mycotoxins: A review. *Mol. Nutr. Food Res.* 2013, 57: 165-186.

[43] Streit, E., Schwab, C., Sulyok, M., et al. Multi-mycotoxin screening reveals the occurrence of 139 different secondary metabolites in feed and feed ingredients. *Toxins* 2013, 5: 504-523.

[44] Goryacheva, I. Y., de Saeger, S., Immunochemical detection of masked mycotoxins: a short review. *World Mycotoxin J.* 2012, 5: 281-287.

[45] Kerr, M. A. and Thorpe, R. (ed.); *Immunochemistry LabFax*. 1994 ßIOS Scientific Publishers, Oxford.

[46] Paul, W. E. (ed.); *Fundamental Immunology* 7th edition. 2013 Wolters Kluwer, Phildelphia.

[47] Shetty, N. (ed.); *Immunology Introductory Textbook*. 2005 New Age International Publishers, Delhi.

[48] Jaypee, N. (ed.); *Textbook of Immunology*. 2014 Jaypee Brothers Medical Publishers Ltd., New Delhi.

[49] Harlow, E. and Lane, D., *Antibodies: A Laboratory Manual*. 1998 Cold Spring Harbor Laboratory Press, Cold Spring Harbor.

[50] Howard, G. C. (ed.); *Basic Methods in Antibody Production and Characterization*. 2009 CRC Press, Boca Rota.

[51] Greenfield, E. A. (ed.); *Antibodies: A laboratory Manual*. 2013 Cold Spring Harbor Laboratory Press, Cold Spring Harbor.

[52] Meulenberg, E. P. (ed.) *Antibodies: Applications and New Developments*. 2012 Bentham Scientific Publications eBooks, Sharjah.

[53] Urusov, A. E., Zherdev, A. V., Dzantiev, B. B., Immunochemical methods of mycotoxin analysis. *Appl. Biochem. Microbiol.* 2010, 46: 253-266.

[54] Meulenberg, E. P., Immunochemical methods for ochratoxin A detection: A review. *Toxins* 2012, 4: 244-266.

[55] Kavanagh, O., Elliott, C. T., Campbell, K., Progress in the development of immunoanalytical methods incorporating recombinant antibodies to

small molecular weight biotoxins. *Anal. Bioanal. Chem.* 2015, 407: 2749-2770.

[56] Fukal, L., Reisnerova, H., Monitoring of aflatoxins and ochratoxin A in Czechoslovak human sera by immunoassay. *Bull. Environ. Contam. Toxicol.* 1990, 44: 345-349.

[57] Fukal, L., A survey of cereals, cereal products, feedstuffs and porcine kidneys for ochratoxin A by radioimmunoassay. *Food Addit. Contamin.* 1990, 7: 253-258.

[58] Rousseau, D. M., Slegers, G. A., van Peteghem, C. H., Radioimmunoassay of ochratoxin A in barley. *Appl. Environ. Microbiol.* 1985, 50, 529-531.

[59] Rousseau, D. M., Slegers, G. A., van Peteghem, C. H., Solid-phase radioimmunoassay of ochratoxin A in serum. *J. Agric. Food Chem.* 1986, 34: 862-865.

[60] Giesen, C., Jakubowski, N., Panne, U., Weller, M. G., Comparison of ICP-MS and photometric detection of an ELISA for the determination of ochratoxin A in wine. *J. Anal. At. Spectrom.* 2010, 25: 1567-1572.

[61] Goryacheva, I. Y., Rusanova, T. U., Burmistrova, N. A., de Saeger, S., Immunochemical methods for the determination of mycotoxins. *J. Anal. Chem.* 2009, 64: 768-785.

[62] Huybrechts, B., Tangni, E., Evaluation of immunoassy kits for ochratoxin A determination in cereals. *AFSCA-FAVV CODA-CERVA*, Tervurnen, Belgium, 2010, pp. 67.

[63] Bazin, I., Nabais, E., Lopez-Ferber, M., Rapid visual tests: Fast and reliable detection of ochratoxin A. *Toxins* 2010, 2: 2230-2241.

[64] Posthuma-Trumpie, G. A., van Amerongen, A., Lateral flow assays. In: *Antibodies: Applications and new developments*. Bentham eBooks, E. P. Meulenberg (ed.), p. 175-183.

[65] Prieto-Simon, B., Campas, M., Immunochemical tools for mycotoxin detection in food. *Monatsh. Chem.* 2009, 140: 915-920.

[66] Anfossi, L., D'Arco, G., Baggiano, C., et al. A lateral flow immunoassay for measuring ochratoxin A: Development of a single system for maize, wheat and durum wheat. *Food Control* 2011, 22: 1965-1970.

[67] Goryacheva, I. Y., deSaeger, S., Nesterenko, I. S., et al. Rapid all-in-one three-step immunoassay for non-instrumental detection of ochratoxin A in high-coloured herbs and spices. *Talanta* 2007, 72: 1230-1234.

[68] Saeger de, S., van Peteghem, C., Flow-through membrane-based enzyme immunoassay for rapid detection of ochratoxin A in wheat. *J. Food Prot.* 1999, 62: 65-69.

[69] Rusanova, T. Y., Beloglazova, N. V., Goryachevva, I. Y., et al. Non-instrumental immunochemical tests for rapid ochratoxin A detection in red wine. *Anal. Chim. Acta* 2009, 653: 97-102.

[70] Huang, B., Xiao, H., Zhang, J., et al. Dual-label time-resolved fluoroimmunoassay for simultaneous detection of aflatoxin A and ochratoxin A. *Arch. Toxicol.* 2009, 83: 619-624.

[71] Maragos, C. M., Fluorescence polarization for mycotoxin determination. *Mycotoxin Res.* 2006, 2: 96-99.

[72] Maragos, C., Fluorescence polarization immunoassay for mycotoxins. A review. *Toxins* 2009, 1: 196-207.

[73] Shim, W-B., Kolosova, A. Y., Kim, Y-J., et al. Fluorescence polarization immunoassay base don monoclonal antibody for the detection of ochratoxin A. *Int. J. Food Sci. Technol.* 2004, 39: 829-837.

[74] Zezza, F., Longobardi, F., Pascale, M., et al. Fluorescence polarization immunoassay for rapid screening of ochratoxin A in red wine. *Anal. Bioanal. Chem.* 2009, 395: 1317-1323.

[75] Taihua, L., Jeon, K-S., Suh, Y. D., Kim, M-G., A label-free, direct and competitive FRET immunoassay for ochratoxin A base don intrinsic fluorescence of an antigen and antibody complex. *Chem. Commun.* 2011, 47: 9098-9100.

[76] Li, T., Byun, J-Y., Kim, B. B., Shin, Y-B., Kim, M-G., Label-free homogeneous FRET immunoassay for the detection of mycotoxins that utilizes quenching of the intrinsic fluorescence of antibodies. *Biosensors Bioelectr.* 2013, 42: 403-408.

[77] Goryacheva, I. Y., de Saeger, S., Eremin, S. A., van Peteghem, C., Immunochemical methods for rapid detection: evolution from single to multiple analyte screening. A review. *Food Addit. Contam.* 2007, 24: 1169-1183.

[78] Goryacheva, I. Y., de Saeger, S., Delmulle, B., et al. Simultaneous non-instrumental detection of aflatoxin B1 and ochratoxin A using a clean-up tandem immunoassay column. *Anal. Chim. Acta* 2007, 590: 118-124.

[79] Anderson, G. P., Kowtha, V. A., Taitt, C. R., Detection of fumonisin B1 and ochratoxin A in grain products using microsphere-based fluid array immunoassays. *Toxins* 2010, 2: 297-309.

[80] Peters, J., Bienenmann-Ploum, M., de Rijk, T., Haasnoot, W., Development of a multiplex flow cytometric microsphere immunoassay for mycotoxons and evaluation of its application. *Mycotox. Res.* 2011, 27: 63-72.

[81] Aqai, P., Peters, J., Gerssen, A., et al. Immunomagnetic microbeads for screening with flow cytometry and identification with nano-liquid chromatography mass spectrometry of ochratoxin in wheat and cereal. *Anal. Bioanal. Chem.* 2011, 400: 3085-3096.
[82] Czeh, A., Mandy, F., Feher-Toth, S., et al. A flow cytometry based competitive fluorescent microsphere immunoassay (CFIA) system for detecting up to six mycotoxins. *J. Immunol. Meth.* 2012, 384: 71-80.
[83] Urraca, J. L., Marazuela, M. D., Merino, E. R., et al. Molecularly imprinted polymers with a streamlined mimic for zearalenone analysis. *J. Chromat. A* 2006, 1116: 127-134.
[84] Yu, J. C. C., Lai, E. P. C., Molecularly imprinted polymers for ochratoxin A extraction and analysis. *Toxins* 2010, 2: 1536-1553.
[85] Mairal, T., Özalp, V. C., Sánchez, P. L., et al. Aptamers: Molecular tools for analytical applications. *Anal. Bioanal. Chem.* 2008, 390: 989-1007.
[86] Zhou, J., Battig, M. R., Wang, Y., Aptamer-based molecular recognition for biosensor development. *Anal. Bioanal. Chem.* 2010, 398: 2471-2480.
[87] Song, K-M., Lee, S., Ban, C., Aptamers and their biological applications. *Sensors* 2012, 12: 612-631.
[88] Rhouati, A., Yang, C., Hayat, A., Marty, J-L., Aptamers: A promising tool for ochratoxin A detection in food analysis. *Toxins* 2013, 5: 1988-2008.
[89] Cruz-Aguado, J. A., Penner, G., Determination of ochratoxin A with a DNA aptamer. *J. Agric. Food Chem.* 2008, 56: 10456-10461.
[90] Barthelmebs, L., Jonca, J., Hayat, A., et al. Enzyme-linked aptamer assays (ELAAs), based on a competitive format for a rapid and sensitive detection of ochratoxin A in wine. *Food Control* 2011, 22: 737-743.
[91] Yang, C., Lates, V., Prieto-Simon, B., et al. Rapid high-throughput analysis of ochratoxin A by self-assembly of DNAzyme-aptamer conjugates in wine. *Talanta* 2013, 116: 520-526.
[92] McKeague, M., DeRosa, M. C., Challenges and opportunities for small molecule aptamer development. *J. Nucl. Acids* 2012, Article ID 748913, pp. 20.
[93] Wang, R. E., Zhang, Y., Cai, J., et al. Aptamer-based fluorescent biosensors. *Curr. Med. Chem.* 2011, 18: 1-10.
[94] Ermolaeva, E. N., Kalmykova, E. N., Piezoelectric immunosensors: analytical potentials and outlooks. *Russ. Chem. Rev.* 2006, 75: 397-409.
[95] Ermolaeva, T. N., Kalmykova, E. N., Shashkanova, O. Y., Piezoquartz biosensors for the analysis of environmental objects, foodstuff and for clinical diagnostic. *Russ. J. Gen. Chem.* 2008, 78: 2430-2444.

[96] Adrian, J., Fernández, F., Muriano, A., et al. Biosensors for pharmaceuticals and emerging contaminants based on novel micro and nanotechnology approaches. *Hdb Env. Chem.* 2009, 5J: 47-68.

[97] Holford, T. R. J., Davis, F., Higson, S. P. J., Recent trends in antibody based sensors. *Biosensors Bioelectr.* 2012, 34: 12-24.

[98] Kimmel, D. W. LeBlanc, G., Meschievitz, M. E., Cliffel, D. E., Electrochemical sensors and biosensors. *Anal. Chem.* 2012, 84: 685-707.

[99] Mariani, S., Minunni, M., Surface Plasmon resonance applications in clinical analysis. *Anal. Bioanal. Chem.* 2014, 406: 2303-2323.

[100] Albers, W. M., Vikholm, I., Viitala, T., Peltonen, J.; Interfacial and materials aspects of the immobilization of biomolecules onto solid surfaces. In: *Handbook of Surfaces and Interfaces of Materials.* H. S. Nalwa (ed.) Vol. 5: Biomolecules, Biointerfaces and Applications. 2001, Academic Press., p. 1-31.

[101] Ngundi, M. M., Shriver, L. C., Moore, M. H., et al. Multiplexed detection of mycotoxins in foods with a regenerable array. *J. Food Prot.* 2006, 69: 3047-3051.

[102] Fu, X-H., Surface plasmon resonance immunoassay for ochratoxin A based on nanogold hollow balls with dendritic surface. *Anal. Lett.* 2007, 40: 2641-2652.

[103] Liu, X. P., Deng, Y. J., Jin, X. Y., et al. Ultrasensitive electrochemical immunosensors for ochratoxin A using gold colloid-mediated hapten immobilization. *Anal. Biochem.* 2009, 389, 63-68.

[104] Yuan, J., Deng, D., Lauren, D. R., et al. Surface plasmon resonance biosensor for the detection of ochratoxin A in cereals and beverages. *Anal. Chim. Acta* 2009, 656: 63-71.

[105] Radi, A. E., Munoz-Berbel, X., Cortina-Puig, M., Marty, J. L., An elelectrochemical immunosensor for ochratoxin A based on immobilization of antibodies on diazonium-functionalized gold electrode. *Electrochim. Acta* 2009, 54: 2180-2184.

[106] Zamfir, L. G., Geana, I., Bourigua, S., et al. Highly sensitive label-free immunosensors for ochratoxin A based on functionalized magnetic nanoparticles and EIS/SPR detection. *Sensors Actuat.* 2011, 159: 178-184.

[107] Dorokhin, D., Haasnoot, W., Franssen, M. C. R., et al. Imaging surface plasmon resonance for multiplex microassay sensing of mycotoxins. *Anal. Bioanal. Chem.* 2011, 400: 3005-3011.

[108] Gaag van der, B., Spath, S., Dietrich, H., et al. Biosensors and multiple mycotoxin analysis. *Food Control* 2003, 14: 251-254.

[109] Palchetti, I., Mascini, M., Electrochemical nanomaterial-based nucleic acid aptasensors. *Anal. Bioanal. Chem.* 2012, 402: 3103-3114.

[110] Tong, P., Zhang, L., Xu, J. J., Chen, H. Y., Simply amplified eletrochemical aptasensor of ochratoxin A based on exonuclease-catalyzed target recycling. *Biosensors Bioelectr.* 2011, 29: 97-101.

[111] Bonel, L., Vidal, J. C., Duato, P., Castillo, J. R., An electrochemical competitive biosensor for ochratoxin A based on a DNA biotinylated aptmer. *Biosensors Bioelectr.* 2011, 26: 3254-3259.

[112] Anklam, E., Stroka, J., Boenke, A., Acceptance of analytical methods for implementation of EU legislation with a focus on mycotoxins. *Food Control* 2002, 13: 173-183.

[113] Zinedine, A., Manes, J., Occurrence and legislation of mycotoxin in food and feed in Morocco. *Food Control* 2009, 20: 334-344.

[114] Berthiller, F., Crews, C., Dall'Asta et al. Masked mycotoxins: A review. *Mol. Nutr. Food Res.* 2013, 57: 165-186.

[115] Binder, E. M., Managing the risk of mycotoxins in modern feed production. *Anim. Feed Sci. Technol.* 2007, 133: 149-166.

[116] Mateo, R., Medina, A., Mateo, E. M., et al. An overview of ochratoxin A in beer and wine. *Int. J. Food Microbiol.* 2007, 119: 79-83.

[117] Kolosova, A., Stroka, J., Evaluation of the effect of mycotoxin binders in animal feed on the analytical performance of standardised methods for the determination of mycotoxins in feed. *Food Addit. Contam.* 2012, 29: 1959-1971.

[118] Aziz, I. G., Rahimi, K., Shateri, S., Ochratoxin: Contamination and toxicity (A review). *Global Veter.* 2012, 8: 519-524.

[119] Friedman, M., Rasooly, R., Review of the inhibition of biological activities of food-related selected toxins by natural compounds. *Toxins* 2012, 5: 743-775.

[120] Sardinas, N., Gil-Serna, J., Santos, L., et al. Detection of potentially mycotoxigenic *Aspergillus* species in *Capsicum* powder by a highly sensitive PCR-based detection method. *Food Control* 2011, 22: 1-4.

[121] Garcia-Cela, E., Ramos, A. J., Sanchis, V., Marin, S., Risk management toward food safety objective achievement regarding to mycotoxins in pistachio: The sampling and measurement uncertainty issue. *Food Control* 2013, 31: 392-402.

In: Ochratoxins
Editor: Daniel Porter
ISBN: 978-1-63483-693-7

Chapter 2

Ochratoxin A (OTA) and Public Health

Daniel Lerda, PhD
Laboratorio de Genética Molecular, Facultad de Medicina,
Universidad Católica de Córdoba, Córdoba, Argentina

Abstract

Contamination of food and agricultural products of various types due to toxigenic mold is a serious problem. Fungi are present naturally and can contaminate food crops under favorable conditions of temperature, relative humidity, pH, nutrients availability and oxygen. Ochratoxin A (OTA) has been proved to have hepatotoxic, carcinogenic, immunosuppressant and teratogenic properties. It has been classified as possibly carcinogenic for humans (Group 2B) and affects the kidney in animals exposed to natural levels of contaminated food. Ochratoxin (OTA) is a mycotoxin that contaminates a wide variety of foods such as corn, barley, green coffee, wine, beer, cocoa, grapes, spices and dried fruits. Plants cells (*Allium cepa*) are very sensitive to the mycotoxin OTA, as was observed at the highest concentration. Under these conditions OTA produced toxicity and cytogenetic injury. Human exposure to mycotoxins is mostly through plant foods. However, additional potential exposure may be through animal foods such as milk, cheese and meat from animals fed with mycotoxin contaminated foods. These can produce toxic waste and biotransformation products. However, while it may not be possible to attribute deaths or illnesses OTA does not mean they have no effect. In fact, in some regions of the world exposure to OTA could be higher and some populations more sensitive to the toxicological effects. Additionally levels of OTA in food frequently exceed regulatory limits. OTA monitoring and action to remove non-compliant food products from the market is a financial burden equally to the health authorities and food businesses. Therefore, there is potential for both public health and economic benefits of reducing OTA contamination in food.

INTRODUCTION

Contamination of food and agricultural products of various types due to toxigenic mold is a serious problem. Fungi are present naturally and can contaminate food crops under favorable conditions of temperature, relative humidity, pH, nutrients availability and oxygen. Additionally, the loss in crops and food contamination by mycotoxins is a major problem since it has been associated with a wide range of negative effects on human health such as carcinogenic, immunotoxic, teratogenic, hepatotoxic and nephrotoxic effects. In particular, the species belonging to the genera *Aspergillus*, *Fusarium* and *Penicillium* are associated with the production of aflatoxin, fumonisin, ochratoxin, patulin, trichothecenes and zearalenone. Ochratoxin A (OTA) has been proved to have hepatotoxic, carcinogenic, immunosuppressant and teratogenic properties. It has been classified as possibly carcinogenic for humans (Group 2B) [1, 2] and affects the kidney in animals exposed to natural levels of contaminated food.

The temperature, humidity and rainfall records contributed directly o indirectly to the proliferation/contamination of the different substratos with micotoxin.

Ochratoxin (OTA) is a mycotoxin that contaminates a wide variety of foods such as corn, barley, green coffee, wine, beer, cocoa, grapes, spices and dried fruits. In tropical areas OTA is produced in coffee beans mainly by *A. ochraceus* and *A. westerdijkiae*. The natural appearance of OTA in green coffee beans has been reported since 1974 [3].

In cattle, the new conditions of intensive management and use of subproducts in the industry enable fungal development and the use of increasingly concentrated diets modify cattle rumen microflora altering the metabolism of mycotoxins and making cattle more susceptible to them.

Human exposure to mycotoxins is mostly through plant foods. However, additional potential exposure may be through animal foods such as milk, cheese and meat from animals fed with mycotoxin contaminated foods. These can produce toxic waste and biotransformation products.

The results of currently available *in vitro* genotoxicity assays do not fully support the assumption that OTA is a genotoxin: negative results were found in DNA repair assays using primary rat hepatocytes (4) and in *Salmonella* microsome assays with S9 mix [4, 5, 6, 7]. Evidence for a direct genotoxic effect was reported in (i) unscheduled DNA synthesis with rat hepatocytes and porcine urinary bladder epithelial cells [8]; (ii) chromosomal aberrations were observed in human lymphocytes exposed to OTA [9]; (iii) micronucleus formation in ovine seminal vesicle cell cultures [9]; and (iv) micronucleus formation in Syrian hamster embryo fibroblasts [9]. In recent investigations, clastogenic activity in humanderived hepatoma (Hep G2) cells was found [10] and DNA damage and cytogenetic effects in rat cells were reported [11].

The common onion *(Allium cepa)* is an excellent plant for the assay of chromosome aberrations that occur after chemical treatment. Protocols were developed using root tips from either bulbs or seeds of *A. cepa* to study cytological endpoints, such as chromosome breaks and exchanges, which follow the testing of chemicals in somatic cells. In this background, and since OTA has not yet been evaluated in plant systems, it was of interest to study toxicological aspects of OTA in relation to plant cells. The aim of the present study was to investigate the genotoxic effects of OTA in *A. cepa*. Since OTA has not yet been evaluated in plants systems, the research focused on describing the controversial effects that OTA when *A. cepa* is used as a vegetal model to demonstrate sensitivity in genotoxin screening.

OTA Formation

The ideal conditions for the growth of different fungal species were generated: *humidity*, the more moisture, the more proliferation. *Temperature*, the temperature range for optimum fungi growth is from 25°C to 35°C, with species that are tolerant up to 55°C and 4°C minimum. We should also keep in mind that fungal stress processes favors their OTA production. Climate change is projected to increase the biomass of crops. Also, alternative host plants may further increase inoculum production. An increase in atmospheric CO2 concentration has been shown to directly increase the amounts of Fusarium Head Blight (FHB) and Crown Rot (CR) inoculum. A deterioration of grain quality may occur as a direct effect of increasing temperature and CO2 which may favour mold growth. Water availability is expected to be altered by climate change as some areas experience drought and others, greater precipitation. The effects of humidity on mycotoxin production in crops are less clear than those for temperature. In a post-harvest setting, an increase in condensation can lead to spoilage which may increase contamination with mycotoxins such as ochratoxin A, aflatoxins and perhaps trichothecenes in damp grain. The distributions and types of mycotoxins post-harvest might change significantly, giving way to new, emerging mycotoxins, their toxicities and possible synergistic interactions with mycotoxins present.

OCHRATOXIN A TOXICITY AND GENOTOXICITY[1]

Onion bulbs of the Pearl variety, average weight 20g, were used. Adventitious roots were obtained by placing the base of bulbs in filtered water in glass tubes equipped with a constant air bubbling system (10–20 $ml.min^{-1}$) in an incubator at 25 ± 0.5 _C in darkness. Control bulbs were incubated in filtered water. Pure, crystalline OTA solutions (Sigma Chemical Co, St Louis, MO, USA) were prepared at concentrations of 0.25, 0.50, 1.0, 2.0, 5.0 and 10 $ug.ml^{-1}$ in (DMSO) and diluted with filtered water. Water was used as the negative control. In order to determine the concentrations necessary for the genotoxicity test, root growth was monitored over 72 h. Between 20 and 30 roots were used at each concentration of OTA, and the average of root lengths were measured and used for calculation of percentage growth of exposed roots with respect to the control. These values were used to produce a dose – response curve. The OTA concentration of 10 $ug.ml^{-1}$ was eliminated from the genotoxicity analysis [12]. All preparations were administered when roots were 2–3-cm long. The effect of different OTA concentrations on longitudinal growth roots was analysed (Figure 1). OTA concentrations of 10 $ug.ml^{-1}$ arrested growth after 24 h, without apparent root death. At concentrations of 0.25, 0.50, 1.0, 2.0 and 5.0 $ug.ml^{-1}$, root growth rate was reduced compared to control roots. These findings indicate that OTA causes inhibition of root growth in a concentration-dependent manner.

In terms of cell proliferation, the frequency of mitotic cells progressively decreased with increasing OTA concentration (Figure 2), reaching a minimum after 24 h.

[1] Lerda, D et al. Allium cepa as a biomonitor of ochratoxin A toxicity and genotoxicity. Plant Biology. 2010, 12:685-688.

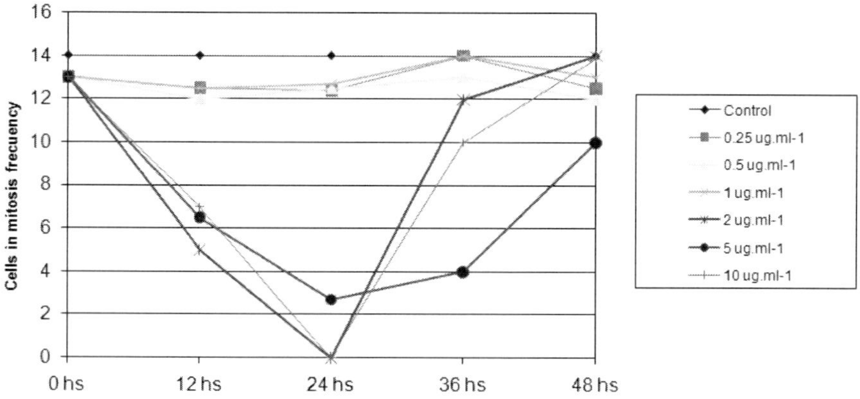

Figure 1. Effects of OTA on onion root length.

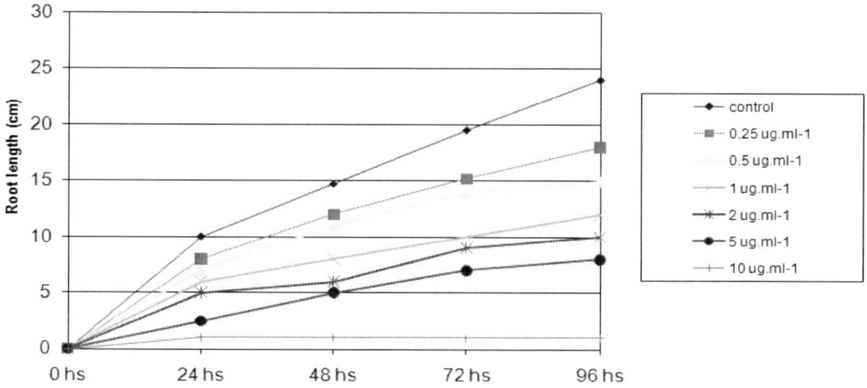

Figure 2. Effects of OTA on proliferation activity in the meristematic zone of onion roots.

This inhibition was transient, as recovery of proliferation activity was observed after 48 h of OTA incubation. These findings suggest that OTA blocks the cell division cycle at a stage before mitosis. Onion roots exposed to 0.25, 0.50, 1.0 and 2.0 ug.ml^{-1} OTA had similar frequency of mitotic cells to control roots. The chromosomal aberrations (CA) were studied with OTA concentrations of 0.25, 0.50, 1.0, 2.0 and 5.0 ug.ml^{-1}, since 10.0 ug.ml^{-1} stopped growth after 24 h. OTA induced clastogenic effects in meristematic roots of Allium cepa. Data on the frequency of aberrant cells are presented in Table 1. An increase in the frequency of aberrant cells was observed with the highest OTA concentration (5.0 ug.ml^{-1}). The Irwin– Fischer (Z) test for exact probability was carried out using the total number of aberrant cells, since 5.0 $ug.ml^{-1}$ OTA only ruptured cells and produced binucleate cells that differed from those of the positive control. The Irwin–Fischer Z was 7.01 (> 1.90) and significant at $P < 0.05$. The detected CA included ruptured chromatids and binucleate cells among other cell types.

Allium cepa offers a good experimental model for in vivo evaluation of genotoxicity of substances and complex mixtures. Although OTA toxicity has been evaluated in several experiments in in vivo and in vitro prokaryotic and eukaryotic cells [13], there are no reports of the effects of this compound in plant cells. Ours findings suggest that OTA inhibits longitudinal growth of the root as a result of the concentration applied. The toxic effect of root growth inhibition can be due to OTA interference with processes associated with root elongation in the zone contiguous to the meristem. It has been known for several decades that different A. cepa root zones have different electronegativity from the apex to the base. Oxygen consumption associated with respiration is higher in the meristematic zone, where cell division is most active [14]. Considering that the reduction of the OTA depends on oxide reduction conditions, as previously demonstrated, it is plausible that the reduction occurs in the elongation zone, where oxidation reactions are less favourable than in the meristematic zone. This could lead to a toxic effect of OTA and growth inhibition of the root, demonstrating that OTA acts in elongation zone and causes toxicity in the meristematic zone. Recent publications attribute to the regulation of elongation and differentiation by the root of A. cepa to differences in ascorbate content and enzymes related to the oxidation state, as well as to presence and activity of hydrogen peroxide and the peroxide ion. The content of these metabolites and enzymes varies in different zones of the root, depending on the degree of cellular differentiation, physiological state and cell metabolic requirements [15]. OTA blocks cell division at a stage before mitosis.

It was also found that onion roots possess OTA elimination or neutralization mechanisms. Based on the results observed in the A. cepa test, we suggest that OTA is a chemical with clastogenic activity. When analysing our results, we found a high incidence of effects of OTA on meristematic cells that had compromised morphology. These cells, when observed in an optical microscope, seemed to be unstructured, with a fragile cellular envelope, increased cell volume and, in several cases, rupture of the cellular membrane. These observations are indicative of cell death [16], where an increase in cell volume leads to rupture of the plasma membrane, exposing the intracellular contents to the external environment.

Table 1. Frequency and spectrum of cytological aberrations induced by OTA in the Allium cepa assay system

Total number	Abnormalities								Frequency of aberrant cells (%)based on
Treatment of cells	Cells in division	breaks	bridges	stickiness	Disturbed meta/ana	bin cells	Total aberrant cell	Total cell scored	Number dividing cell
0.25 ug.ml-1 4500	359	–	–	–	–	–	–	–	–
0.50 ug.ml-1 4500	360	–	–	–	–	–	–	–	–
1.0 ug.ml-1 4500	370	–	–	–	–	–	–	–	–
2.0 ug.ml-1 4500	344	–	–	–	–	–	–	–	–
5.0 ug.ml-1 4500	370	6	–	–	-	5	11*	0.26	3.1
Negative control 5000	366	–	–	–	–	–	–	–	–
Positive control (a) 5000	361	7	8	18	7	4	44	0.99	11.0

–: No clastogenic effects.

*Z = 7.01 significant at $P < 0.05$ compared with negative and positive controls.

a: Ethyl methyl sulfoxide (0.2%).

Cytologic changes observed in the Allium chromosome aberrations assay, at concentrations of 5.0 $ug.ml^{-1}$ OTA, showed that OTA was able to induce genotoxicity at the chromosome level. The A. cepa assay has excellent correlations with mammalian systems [17]. Results obtained on Allium Chromosomal Aberrations (CA) may be used as a relevant test for the general detection of genotoxin in the environment [12, 17, 18, 19, 20, 21, 22, 23, 24, 25].

There are no studies reporting OTA adverse effects on plants cells. The present results indicate that plant cells (Allium cepa) have a very sensitive cellular response to the mycotoxin, OTA, at the highest concentration used. This concentration caused toxicity and cytogenetic injury. Evidence in vitro and in vivo indicates that OTA can damage plant DNA indirectly.

OTA in Plant Food[1]

The natural appearance of OTA in green coffee beans has been reported since 1974 [3]. Gopinandhan and co-workers [26] found OTA rates between 0.2 and 13.5 $ng.g^{-1}$ in green coffee samples. The presence of OTA in roasted and prepared coffee was reported by Tsubouchi and co-workers [27]. Before these reports, it was generally accepted that OTA was eliminated during roasting; nevertheless, concentrations above 20 $ng.g^{-1}$ have been reported in commercialized roasted coffee [28]. Several reports concerning the impact of roasting on the OTA content in coffee have shown OTA reduction ranging from 0 – 12% to 90 – 100% [29]. Such variations may be related to the different analytical conditions, on the roasting process or to the heterogeneity in the toxin distribution [30].

We use the current knowledge on OTA degradation during coffee roasting and then to apply the *Allium cepa* genotoxicity test on the roasting-products, using contaminated and non-contaminated coffee beans.

Green *Coffea Arabica* beans resulting from drying and de-husking of coffee fruits, were harvests from Sao Paulo State, Brazil, acquired directly from Terra Forte exporter, Sao Joao Da Boa Vista, SP, Brazil. Green coffee beans was artificially contaminated with *ochratoxin A*.

[1] Lerda, D. et al. Roasting coffee beans (Coffea Arabica) artificially contaminated with ochratoxin A strongly reduces the analytical ochratoxin A content but not the genotoxic effects. Current Topics in Toxicology. 9: 75 -80, 2013.

Green coffee and roasted coffee were analyzed for OTA content [31]. Green coffee beans without OTA was used as control.

In the present study, we observed that the roasting process produces a substantial OTA reduction; the remaining compounds can combine with those of coffee and produce some toxicity in plant cell. This is probably due to intrinsic components such as caffeic acid and other chlorogenic acids, as well as hydrogen peroxide. String electrophiles generated through Fenton type chemistry then react with DNA and other cell components, and measurable change is surely visible. The most interesting finding in the present study was that coffee beans non-contaminated with OTA and roasted at 230° C for 120 minutes showed clastogenic and mutagenic effects. This reveals that some coffee cell compounds change during the roasting process producing several compounds that have harmful effects on plant cells. Ames and Gold [32] identified more than 20 rodent carcinogens in coffee several of them linked to chlorogenic acids, aldehydes and other components. In the present study, it was observed that the highest OTA concentrations in green and roasted coffee showed chromosomal aberrations and the con-contaminated coffee sample showed a concentration-dependent reduction in the mitotic index and most of the common abnormalities were the bridges. Bridges probably occur by the interruption and joining of chromosome or chromatids [33], or as a result of chromosome stickiness, or it could be attributed to unequal translocation or inversion of chromosome segments [34]. Cytologic changes observed in the *Allium* chromosome aberrations assay showed that OTA was able to induce genotoxicity at the chromosome level. The efficacy of the micronucleus test for mutagenicity screening has been well established. Micronuclei are a manifestation of the chromosome breakages and the failure of the normal spindle function. It is likely that MN formation is due to clastogenic effects.

Genotoxicity analyses on meristematic cells of *A. cepa* indicated that the roasting process was not efficient enough for OTA degradation because the clastogenic and mutagenic effects were not reduced. This shows that OTA-degraded compounds could have possibly combined with *coffea arabica* compounds forming compounds toxic to vegetable cells.

OTA IN ANIMAL FOODS[2]

In general there is no data of OTA poisoning in cattle, due to the ability of the rumen protozoa microflora to metabolize and hydrolyse OTA to ochratoxin-alpha that is not toxic and does not degrade. This ability is affected by the variation of rumen microflora due to the consumption of the different food types that are assigned to cattle. There are estudies in which it is proved that different types of diets have a large influence on the metabolization of some mycotoxins. Thus, a diet based on 100% hay leads the rumen fluid to a pH of 7.1, ochratoxin A is hydrolyzed to ochratoxin alpha (non-toxic) in 36 minutes. When the diet contains 70% hay and 30% concentrate the pH of rumen liquid is 6.5, OTA hydrolysis takes longer (1:18 hours). If diets with 100% concentrate are consumed, this hydrolysis takes 3:18 hours [35]. This could explain the absorption of unhydrolyzed mycotoxins and cause different pathologies.

In our country there are not many precedents on pathologies associates with consumption of food in feedlots, for this reason we consider it useful for feedlots to take the necessary measures to control the portions they use as food.

In a beef cattle feedlot for human consumption, located in San Agustin village, Calamuchita department, Córdoba, Argentina, a marked mortality rate took place during the months of February 2013 to June 2014. Weather conditions were atypical for the area and season; there were high temperatures, droughts followed by periods of excess moisture and rain. The ration animals were fed with was composed of corn, alfalfa hay, corn burlanda, gluten feed, peanut shells and sunflower pellet. Affected animals presented various symptoms such as dyspnea, hemoglobinuria and hematuria, lack of coordination, death and in many cases, sudden death. Bovines underwent necropsy by which jaundice, hepatitis with focal necrosis, gallbladder edema, hemoglobinuria, hematuria and kidney necrosis were found. The content of aflatoxins and ochratoxins (OTA) in the ration was determined, detecting a high amount of OTA which nay have been the cause of cattle mortality.

The different components of the ration have already been mentioned: corn, alfalfa hay, sunflower pellets, corn burlanda, gluten feed. It is important to remember that corn burlanda is s byproduct of the production of ethanol and that it concentrates loads of mycotoxins or fungi that could be brought from

[2] Torassa, D et al. Mycotoxin Poisonong in an Intensive Beef-Fattening Sustyem. Animal and Veterinary Sciences. 2 (6): 233-227, 2014.

the corn from which it originated. Gluten feed is the byproduct of the extraction of glucose from the corn grain. Burlanda and gluten feed have 67% humidity and a high Aw (water activity) favoring the development of different species of fungi [36, 37].

Table 2. Cattle death

	January	February	March	April	May	June	July	August	September	October	November	December
2013		3		4		2	2	1			1	5
2014	11	50	15	2	1	0						

Table 2. Cattle death

	January	February	March	April	May	June	July	August	September	October	November	December
2013		3		4		2	2	1			1	5
2014	11	50	15	2	1	0						

The mycotoxin aflatoxin were alayzed with the Veratox Quantitative Aflatoxin test, ELISA kit, Neogen (USA) and by the Tam et al. HPLC method [38] and ochratoxin were determined with the veratox Quantitative Ochratoxin tes, ELISA kit and by the Ng et al. HPLC method [39]. The clinical signs presented were: sudden death, dyspnea, anorexia, polydipsia, jaundice, lethargy, ambulation, anuria, red urine (hematuria and hemoglobinuria), and there was no increase in body temperature.

Table 3. Content of mycotoxins in different samples

Substratum	Aflatoxin ppb	Ochratoxin ppb
Ration	5.2	29.5
Corn Burlanda	4.3	29.9
Gluten feed	4.1	32.5
Corn	3.3	5.9

Table 2 shows that during 2013 and 2014 cattle deaths occurred on site. It can be seen how they increased during the months of January, February and March 2014.

The values found of mycotoxins in different samples can be in Table 3.

The content of mycotoxins with mortality in the months of January to Abril 2014 are shown in Table 4.

Table 4. Relationship of mycotoxins and monthly mortality

	Aflatoxin / ppb	Ochratoxin / ppb	Number of dead animals
January/14	8.1	7.3	11
February/14	5.0	29.5	50
March/14	3.3	8.2	15
April/14	3	1.3	2

Urinalysis:
Appearance: Cloudy
Color: Reddish
Density: 1018
Ph: 7
Hemoglobin: ++++
Proteins: ++
Bilirubin: none
Urobilinogen: none
Glucose: none
Ketones: none
Nitrite: none
Sediment
Cylinders: none
Thin epithelial cells: none
Round transition Cells: none
Renal and tubular cells: Frequently isolated and grouped 7 / C
Leukocytes: Some isolated and grouped 2 / C
Erythrocytes: Some isolated and grouped 2 / C
Bacteria: none
Crystals: none
Findings at necropsy:

Highly marked Jaundice in mucosa and subcutaneous adipose tissue. Dark intestines, congestive and hemorrhagic mucosa with arborization. Bloody feces. Hemorrhagicenteritis. Gas production is observed.

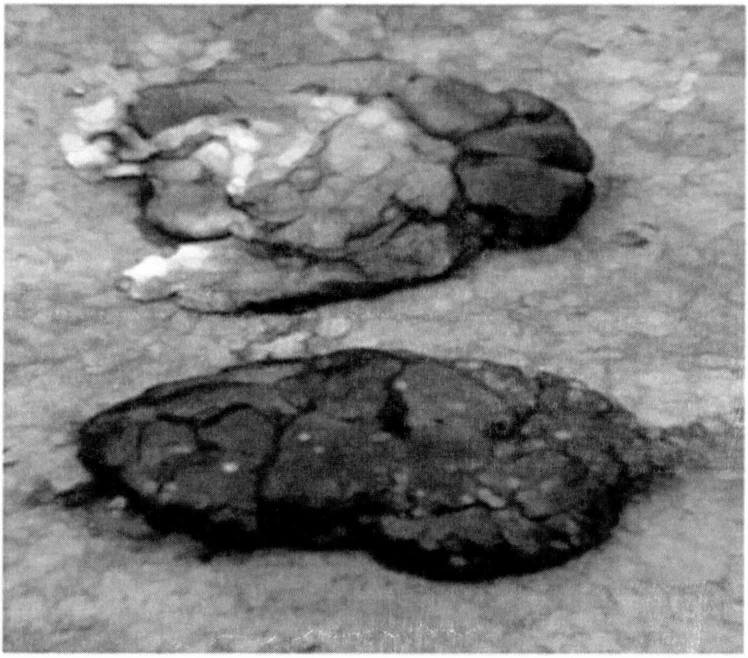

Figure 3. Kidney black, friable, autolytic.

Dark Kidneys from gray to deep black, friable and in some cases amorphous even autolytic (Figure 3) hematuria and hemoglobinuria (not always), full-fledged bladder (anuria) very strong generic smell.

Liver with an increased size (Hepatomegaly), a deep ocher color, areas of necrosis, friable to touch, full-fledged gallbladder with emphysema and edema, very dense umber bile with the lumpy appearance of clay (Figure 4).

In some cases only the liver injury was found.

Muscles have a diminished color (pale pink).

Macroscopically livers with a firm consistency, regular edges and smooth surfaces were observed, some exhibited a pale color, others had a marked lobular pattern and most of them had jaundice of varying intensity. The pieces sent were of various sizes.

Figure 4. Hepatitis, liver ocher gallbladder eventful emphysema perivesicular.

Microscopically the liver parenchyma showed necrosis with centrilobular predominance (Figure 5), varying amounts of fat change, lipidosis, biliary hyperplasia, condensation of fibrous stroma (Figure 6) and eventually regenerating nodules. Hepatocytes showed atypia, vacuolization, degeneration and some had extremely large cores and larger volume (megalocytes) (Figure 7).

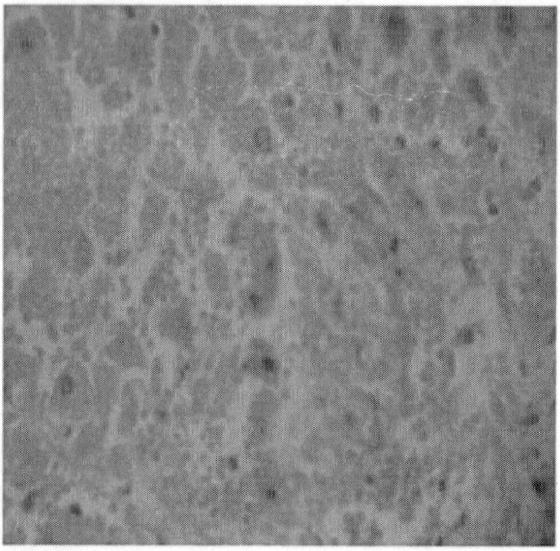

Figure 5. Centrilobular necrosis HE 40 X.

Figure 6. Stromal diffuse fibrosis HE 40 X.

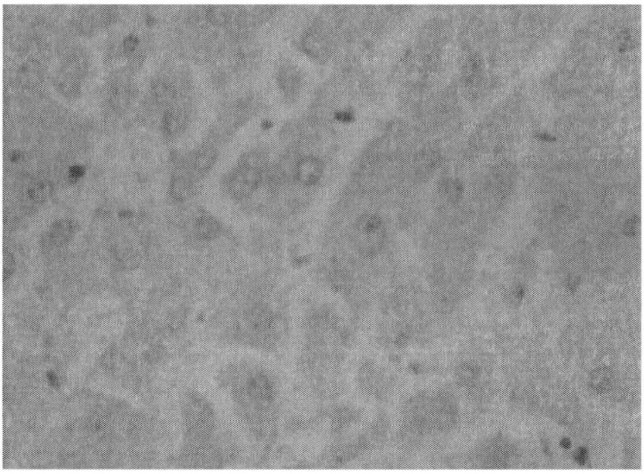

Figure 7. Megalocytes. HE. 40 X.

Macroscopically pieces belonging to the *kidney* were observed, some had a solid consistency, when cut some revealed a semi-soft consistency, with regular edges, smooth surfaces and different measures (3 x 2,2 x 1,6cm; 3 x 2,7x2cm; 2,5 x 2 x 2cm; 4 x 2 x 2cm and 3 x 2 x 1, 7cm)

Microscopically renal parenchyma with interstitial nephritis was observed there was also presence of interstitial, intertubular and interglomerular fibrin. The cells of the renal tubules are degenerated, vacuolated. Diffuse necrosis (Figure 8). Other sections showed congestion with preserved parenchyma and stroma.

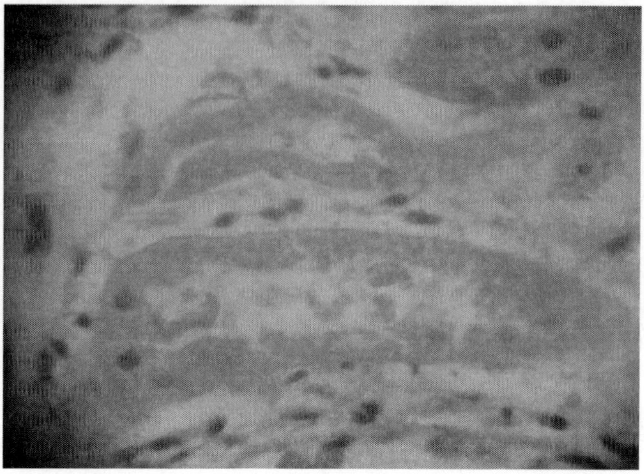

Figure 8. Necrosis diffuse HE 40 X.

These histopathological lesions corresponding to diffuse hepatic fibrosis and renal lesions are not pathognomonic of a particular disease as they occur in different toxins (aflatoxins, OTA, pyrrolizidine alkaloids, nitrosamines) [40, 41] but it is important to accompany visible histopathological lesions with the context, rodeo clinical characteristics and other confirmatory studies.

Diagnosis: It was made from injuries and the confirmation of mycotoxins in the rations given to the animals.

Differential Diagnosis:

Leptosirosis:

Seorología: Martin and Petit Microagglutination was conducted to investigate Leptospira interrogans, serovars: pomona, wolffi, Icterohemorrhagic and canicola. The results were NEGATIVE.

Bacillary hemoglobinuria: No consistent lesions were found.

Although prevention of mycotoxin contamination in the field is the main objective of the agriculture and food industries, methodologies must be developed during the processing of food for slowing the growth of fungi.

Post-harvest strategies aim at the reduction of fungal contamination and, consequently, the levels of mycotoxins in agricultural products during storage, handling, processing and transportation. Such strategies include the improvement of the conditions of drying and storage, the use of chemical and natural agents, and irradiation. Unfortunately, contamination cannot be completely avoided. Therefore, there is an increased focus on efficient detoxification methods for mycotoxins present in foods, and in inhibiting mycotoxin absorption in the gastrointestinal tract. It is also possible to prevent the toxic effects of mycotoxins using feed additives, such as antioxidants, sulfur-containing amino acids, vitamins and trace elements. A new strategy to control mycotoxicosis is the application of microorganisms (lactic acid bacteria, yeast) capable of biotransforming some mycotoxins into less toxic metabolites.

After confirmation of the presence of toxins in the feed ration, the component of the ration that presented a higher load of mycotoxins was identified in order to limit its inclusion or to remove it. Mycotoxin adsorbent was used in the ration [5].

A control plan was implemented through periodic surveys directed at the mycotoxin content of the ration to establish its seasonal incidence and to anticipate the occurrence of poisoning.

In this investigation the levels of mycotoxins in bovine feed was reduced using absorbents, as a result of this cattle mortality in the establishment ceased.

How to Avoid the Formation of OTA

Although prevention of mycotoxin contamination in the field is the main objective of the agriculture and food industries, methodologies must be developed during the processing of food for slowing the growth of fungi.Post-harvest strategies aim at the reduction of fungal contamination and, consequently, the levels of mycotoxins in agricultural products during storage, handling, processing and transportation. Such strategies include the improvement of the conditions of drying and storage, the use of chemical and natural agents, and irradiation. Unfortunately, contamination cannot be completely avoided. Therefore, there is an increased focus on efficient detoxification methods for mycotoxins present in foods, and in inhibiting mycotoxin absorption in the gastrointestinal tract. It is also possible to prevent the toxic effects of mycotoxins using feed additives, such as antioxidants, sulfur-containing amino acids, vitamins and trace elements. A new strategy to control mycotoxicosis is the application of microorganisms (lactic acid bacteria, yeast) capable of biotransforming some mycotoxins into less toxic metabolites.

Public Health

The results of the studies in plant cells OTA (*Allium cepa*) show a very sensitive OTA cellular response at the highest concentration used. This concentration was toxic and cytogenetic damage. The OTA can damage the DNA of the plant indirectly.

Studies on the genotoxicity of *Coffea arabica* roasted and contaminated with OTA in *A. cepa* meristematic cells showed that the roasting process was not efficient enough to degrade OTA because mutagenic and clastogenic unabated. This shows that the OTA-degraded compounds in combination with compounds of *Coffea arabica* form toxic compounds for plant cells.

In a beef cattle feedlot for human consumption a marked mortality rate took place. The ration animals were fed with was composed of corn, alfalfa hay, corn burlanda, gluten feed, peanut shells and sunflower pellet. Affected animals presented various symptoms such as dyspnea, hemoglobinuria and hematuria, lack of coordination, death and in many cases, sudden death. The content of aflatoxins and ochratoxins (OTA) in the ration was determined, detecting a high amount of OTA which nay have been the cause of cattle mortality-In general there is no data of OTA poisoning in cattle, due to the ability of the rumen protozoa microflora to metabolize and hydrolyse OTA to ochratoxin-alpha that is not toxic and does not degrade.

However, while it may not be possible to attribute deaths or illnesses OTA does not mean they have no effect. In fact, in some regions of the world exposure to OTA could be higher and some populations more sensitive to the toxicological effects. Additionally levels of OTA in food frequently exceed regulatory limits. OTA monitoring and action to remove non-compliant food products from the market is a financial burden equally to the health authorities and food businesses. Therefore, there is potential for both public health and economic benefits of reducing OTA contamination in food.

References

[1] IARC (International Agency for Research on Cancer).Some naturally occurring substances: Food items and constituents, heterocyclic aromatic amines and mycotoxins, IARC monographs on the evaluation of carcinogenic risks to humans, *IARC Press*, Lyon. 1993, 56:489-521.

[2] Pfohl-Leszkowicz, A. and Casteganro, M. L`ochratoxine A., A. Pfohl-Leszkowicz, (Ed.) Les Mycotoxines dans l`alimentation: evolution et gestion des risqué (Ed.), Paris. France: *Lavoisier,* 1999, 249-277.

[3] Levi, C. Trenk, H. and Mohr, H.Study of the occurrence of OTA in green coffee beans. *Journal Association of Official Analytical Chemists.* 1974, 57, 866.

[4] Bendele A.M., Neal S.B., Oberly T.J., Thompson C.Z., Bewsey B.J., Hill L.E., Rexroat M.A., Carlton W.W., Probst G.S. Evaluation of ochratoxin A for mutagenicity in a battery of bacterial and mammalian cell assays. *Food Chemical Toxicology* 1985.23: 911–918.

[5] Kuczuk M.H., Benson P.M., Heath H., Hayes A.W. Evaluation of the mutagenic potential of mycotoxins using Salmonella typhimurium and Saccharomyces cerevisiae. *Mutation Research.* 1978, 53: 11–20.

[6] Wehner F.C., Thiel P.G., van Rensburg S.J., Demasius I.P. Mutagenicity to Salmonella typhimurium of some Aspergillus and Penicillium mycotoxins. *Mutation Research.* 1978. 58: 193–203.

[7] Würgler F.E., Friederich U., Schlatter J.Lack of mutagenicity of ochratoxin A and B, citrinin, patulin and cenestine in Salmonella typhimurium TA102. *Mutation Research.* 1991, 261: 209–216.

[8] Dörrenhaus A., Fo¨llmann W. Effects of ochratoxin A on DNA repair in cultures of rat hepatocytes and porcine urinary bladder epithelial cells. *Archives of Toxicology.* 1997, 71: 709–713.

[9] JECFA. Safety evaluation of certain mycotoxins in food. World Health Organization Food Additives Series. *World Health Organization, Geneva, Switzerland.* 2001, Vol. 47

[10] Ehrlich V., Darroudi F., Uhl M., Steinkellner H., Gann M., Majer B.J., Eisenbauer M., Knasmu¨ller S. Genotoxic effects of ochratoxin A in human-derived hepatoma (HepG2) cells. *Food and Chemical Toxicology.* 2002 40: 1085–1090.

[11] Mally A., Pepe G., Ravoori S., Fiore M., Gupta R.C., Dekant W., Mosesso P. Ochratoxin A causes DNA damage and cytogenetic effects but no DNA adducts in rats. *Chemical Research Toxicology.* 2005, 18: 1253–1261.

[12] Fiskesjo G.The Allium test as a standard in environmental monitoring. *Hereditas.* 1985, 102: 99–112.

[13] WHO (World Health Organization). Safety evaluation of certain food additives and contaminants. World Health Organization Food Additives Series, International Programme on Chemical Safety, Geneva, Switzerland. 2008, Vol. 59.

[14] Berry L., Brock M. Polar distribution of respiratory rate in the onion root tip. *Plant Physiology*. 1946, 21: 542–549.

[15] Cordoba-Pedregosa M., Cordoba F., Villalba J., Gonzalez Reyes J. Zonal changes in ascorbate and hydrogen peroxide content, peroxidase and ascorbate-related enzyme activities in onion roots. *Plant Physiology*. 2003, 131: 697–706.

[16] Cristea I., Esposti M. Membrane lipids and cells death: an overview. *Chemical Physiology Lipids*. 2004, 129: 133–160.

[17] Grant W. Chromosome aberration assays in Allium A report of the US Environmental Protection Agency gene tox program. *Mutation Research*. 1982, 99: 273–291.

[18] Grant W. The present status of higher plants bioassays for the detection of environmental mutagens. *Mutation Research*. 1994, 310: 175–185.

[19] Fiskesjo G. The Allium test – an alternative in environmental studies: the relative toxicity of metal ions. *Mutation Research*. 1988, 197: 243–260.

[20] Fiskesjo G.The Allium test in wastewater monitoring. *Enviromental Toxicology and Water Quality*. 1993, 8: 291–298.

[21] Lerda D. The effect of lead on Allium cepa L. *Mutation Research*. 1992, 281, 89–92.

[22] Nielsen M., Rank J. Screening of toxicity and genotoxicity in wastewater by the use of Allium tests. *Hereditas*. 1994. 121: 249–254.

[23] Smaka-Kinel V., Stegnar P., Lovka M., Toman M. The evaluation of waste, surface and ground water quality using the Allium test procedure. *Mutation Research*. 1996, 368: 171–179.

[24] Bolle P., Mastrangelo S., Tucci P., Evandri M. Clastogenicity of atrazine assessed with the Allium cepa test. *Environmental and Molecular Mutagenesis,* 2004. 43: 137–141.

[25] Feretti D., Zerbini A., Zani C., Ceretti E., Moretti M., Monarca S. Allium cepa chromosome aberration and micronucleus tests applied to study genotoxicity of extracts from pesticide treated vegetables and grapes. *Food Additives and Contaminants*. 2007, 24: 561–572.

[26] Gopinandhan, T.N., Kannan, G. S., Panneerselvam, P., Velmourougane, K., Raghuramulu, Y., and Jayarama, Survey on ochratoxin A in Indian green coffee destined for export. *J Food Addit. Contam. Part B*. 2008 1:51-7.

[27] Tsubouchi, H. Yamamoto, K. Hisada, K. Sakabe, Y. and Udagawa, S. Effect of roasting on ochratoxin A level in green coffee beans inoculated with Aspergillus ochraceus. *Mycopathologia* 1987, 97: 111-5.

[28] Mounjouenpou, P. Durand, N. Guyot, B. and Guiraud, J. Effect of operating conditions on ochratoxin A extraction from roasted coffee. *Food Additives & Contaminants*. 2007, 24: 730 - 734.

[29] Anezqueta, S, Gonzalez-Penas, E., Murillo-Arbizu, M and Lopez de Cerain, A. Ochratoxin A decontamination. *A review. Food Control.* 2009, 20: 326-333.

[30] Suarez-Quiroz, M. De Louise, B. Gonzalez-Ríos, O. Barel, M. Guyot, B. Schorr-Galindo, S. and Pierre Guiraud, J. The impact of roasting on the ochratoxin A content of *coffee International Journal of Food Science and Technology*. 2005, 40: 605-611.

[31] Nakajima, A. Horikoshi, T. and Sakaguchi, T Studies on the accumulation of heavy metal elements in biological systems II. Chemical state of copper accumulated in *Chlorella regularis. J. Agric. Chem.* 1977, 51: 507-511.

[32] Ames, B.N. and Gold, L.S. Chemical carcinogenesis: Too many rodent carcinogens *Proc. Natl. Acad. Sci.* 1990, 87: 7772-7776.

[33] Turkoglus, S. Genotoxicity of Five Food Preservatives Tested on Root Tips of Allium cepa L. *Mutation Research,* 2007, 626:4-14.

[34] Gömürgen, A. Cytological effects of the potassium metabisulphite and potassium nitrate food preservative on root tips of *Allium cepa* L. *Cytologia.* 2005,70: 119-128.

[35] Gimeno, G, Martins, M. Mycotoxins and mycotoxicosis in animals and humans. *Special Nutrients, INC.3rd edition,* 2011, 18 – 56.

[36] Mendoza, A, La Manna, A, Mieres, J. Acosta, Y. Deoxynivalenol intake assessment and commercial mycotoxin binder in grazing dairy cows. *Agrociencia;* 2014.Vol 18, N° 1: 133-140.

[37] Giraudo, J. Mycotoxin poisoning in cattle. *Notes for Veterinary Clinicians.* 2011: 1-12.

[38] Tam, J. Mankotia, M. Mably, M. Pantazopoulos, P. Neil, R. Calway, P. Scott, P. Survey of breakfast and infant cereals for aflatoxins B1, B2, G1 and G2. *Food Addit Contam A.* 2006. 23:693-699.

[39] Ng, P. Mankotia, M. Pantazopoulos, P. Neil, R. Scott, P. P-Y Lau, B. Survey of dry pasta for ochratoxin A in Canada. *J. Food Prot.* 2009. 72:890-893.

[40] Pathologic basis of veterinary disease. Fourth edition By M. Donald McGavin, MVSc, PhD, FACVSc and James F. Zachary, DVM, PhD. 2006 :440-446.

[41] Patología de los animales doméesticos KVF Jubb-PC Kennedy.2007. Vol 2: 246- 269.

In: Ochratoxins
Editor: Daniel Porter

ISBN: 978-1-63483-693-7

Chapter 3

OCHRATOXIN A: POSSIBLE MECHANISMS OF TOXICITY

R. K. Asrani[1,*], Vikram Patial[2] and Meenakshi Thakur[1]

[1]Department of Veterinary Pathology,
Dr. G C Negi College of Veterinary and Animal Sciences,
CSK HPKV, Palampur, H.P., India
[2]Regulatory Research Center, CSIR-IHBT, Palampur, H.P., India

ABSTRACT

The chlorophenolic mycotoxin ochratoxin A (OTA) is prevalent in food sources and represents a potential health hazard to various animals. OTA is a potent renal carcinogen in rodents and has been linked to the fatal human kidney disease, Balkan Endemic Nephropathy (BEN). Understanding the mechanism of OTA-mediated toxicity is vital for the assessment of the risk associated with its intake. Two theories have been proposed in support of mechanism of action of OTA toxicity, i.e., direct genotoxic (DNA binding of OTA moiety) and epigenetic or non-genotoxic (without any direct DNA binding of OTA moiety). OTA facilitates guanine-specific DNA adducts *in vitro* in orally dosed rat and pig kidneys. A low level of DNA lesions were observed in OTA treated animals by ^{32}P-post labeling technique suggesting that OTA may be activated into some other reactive metabolites which further react with DNA to form adducts. Oxidative stress also accounts for the OTA-

[*] Corresponding author: Email: asranirk@gmail.com.

mediated toxicity, thus favoring the epigenetic or non-genotoxic mechanism. The mechanisms of ochratoxicity include inhibition of mitochondrial respiration correlated with a depletion of ATP, inhibition of protein synthesis, enhanced lipid peroxidation and alteration of signal transduction. The combination of these basic mechanisms with an unusual long plasma half-life time (35-70 days), the metabolic conversion of OTA into still active derivatives and conjugates, both potentially reactive with cellular biomolecules including DNA could partially explain the multiple toxic effects of OTA including genotoxicity, cytotoxicity and carcinogenicity.

INTRODUCTION

Ochratoxin A (OTA) is a mycotoxin produced as a secondary metabolite by some fungal species of genera *Aspergillus* and *Penicillium* that contaminates a great variety of food stuffs, such as cereals, coffee beans, grapes and dried fruits (Pfohl-Leszkowicz and Manderville 2007; Duarte et al. 2010; Vettorazzi et al. 2014). As a consequence of its widespread occurrence in food, the human population is continuously exposed to OTA (Clark and Snedeker 2006). OTA includes a wide range of detrimental effects, including nephrotoxicity, hepatotoxicity, carcinogenicity, teratogenicity, immunotoxicity and neurotoxicity in animals (Ringot et al. 2006).

In humans, it has been speculated that OTA may be associated with Balkan Endemic Nephropathy (BEN) and with an increased incidence of urinary tract tumors (Castegnaro et al. 2006). BEN is a slowly progressing, chronic, non-inflammatory, bilateral, primarily tubulointerstitial, distinct nephritis, eventually leading to renal failure and death (Polenakovic and Steffen 2001). Urinary tract tumors are found frequently associated with BEN, up to 45% of BEN patients have been reported to have urinary tract tumors (Sostaric and Vukelic 1991).

OTA has been classified as a possible carcinogen to humans (Group 2B) by the International Agency for Research on Cancer (IARC) (IARC, 1993). Moreover, OTA possesses the longest half-life for its elimination as compared to the other mycotoxins (Iha and Trucksess 2010). OTA is chemically stable against heat and during industrial processing (Jorgensen 2005; Ringot et al. 2006). A provisional tolerable weekly intake level of 100 ng OTA/kg body weight has been established by the Joint Food and Agriculture Organization/World Health Organization Expert Committee on Food Additives (JECFA) (JECFA, 2001). Therefore, it becomes important to understand the possible mechanisms of OTA-mediated toxicity, so that one can assess the risk associated with it and restrict its intake.

Structure of OTA

OTA (Figure 1) consists of a *para*-chlorophenolic moiety containing a dihydroiso-coumarin group that is amide-linked to L-phenylalanine (Pfohl-Leszkowicz & Manderville 2007).

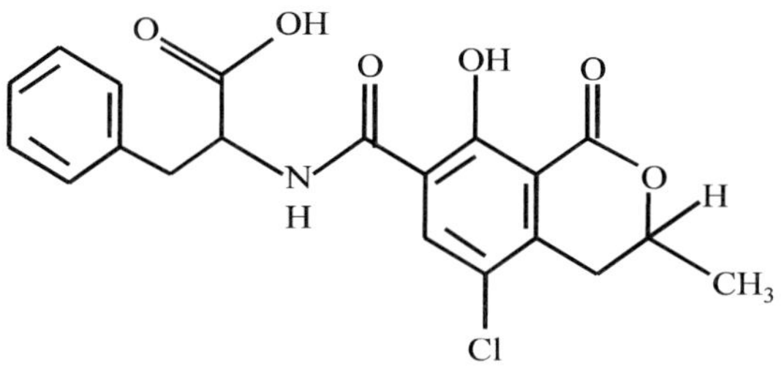

Figure 1. Chemical structure of OTA.

OTA competes with phenylalanine-tRNA synthetase thereby inhibiting the peptide elongation which is an important effect of OTA toxicity. OTA can act on all metabolic systems involving phenylalanine. The phenylalanine moiety of OTA can bind to and inhibit phenylalanine hydroxylase, and thus impair hydroxylation of phenylalanine to tyrosine, which is the key regulatory step in the catabolism of this amino acid (Ringot et al. 2006). The structure-activity studies have suggested that *in vitro* and *in vivo* toxicity of OTA may be attributed to its phenylalanine moiety and *para*-chlorophenolic group (Malaveille et al. 1994; Xiao et al. 1996).

Absorption and Distribution

In animals, OTA is rapidly absorbed after oral ingestion (EFSA, 2006). The absorption of OTA takes place passively through the gastrointestinal tract (Ringot et al. 2006), stomach (Galtier 1977) and proximal jejunum (Kumagai and Aibara 1982). The passive absorption is greatly favored by the high binding affinity of OTA to plasma proteins (Ringot et al. 2006).

In rats and humans, the unbound fraction of OTA is as low as 0.02%, indicating an extent of plasma protein binding of 99.98% (Hagelberg et al. 1989). This binding facilitates its passive absorption in the non-ionized form and partly explains its long-half life in the body (Ringot et al. 2006). Stojkovic et al. (1984) found that in human and porcine serum, OTA binds more specifically to small proteins (molecular mass around 20 kDa) than to albumin. Due to passage of such small proteins through the glomerular membrane, OTA exerts nephrotoxic effect on mammals. OTA accumulates in various organs including kidneys, liver and muscles. However, kidneys are the main target of OTA accumulation. Reabsorption of OTA in proximal and distal tubules of kidneys contributes to the long half-life of OTA and increased nephrotoxicity.

MECHANISMS OF OTA TOXICITY

There is a great controversy regarding the possible mechanisms of OTA toxicity. Two theories have been put forward by researchers in support of its mechanism of action, i.e., direct genotoxic (DNA binding of OTA moiety) and epigenetic or non-genotoxic (without any direct DNA binding of OTA moiety).

One group of researchers argues that OTA undergoes biotransformation to induce genotoxicity. They used ^{32}P-postlabeling method to observe that OTA-derived electrophiles form adducts with DNA. DNA adduction is strongly correlated with carcinogenesis. A second group of researchers report negative findings for the genotoxicity of OTA and argues that OTA does not undergo biotransformation to form electrophilic intermediates capable of reacting with biopolymers. According to this group, oxidative stress rather than genotoxicity accounts for OTA mediated toxicity. However, OTA and its metabolites have been reported to be both mutagenic (Obrecht-Pflumio et al. 1999) and non-mutagenic (Follmann and Lucas 2003) in a range of microbial testing systems.

(i) Direct Genotoxic Mechanism

A ^{32}P-post labeling technique was used by some researchers to find out the genotoxic effects of OTA in animals. They observed low levels of DNA lesions in various organs of OTA treated animals (Pfohl-Leskowicz et al. 1993; Castegnaro et al. 1998) suggesting that OTA may be activated into some other reactive metabolites which may further react with DNA to form adducts

(Faucet et al. 2004; Pfohl-Leskowicz and Castegnaro 2005). OTA facilitates guanine-specific DNA adducts *in vitro* in orally dosed rat and pig kidneys. It has been shown that rat and mice treated with OTA show numerous DNA adducts (Pfohl-Leszkowicz et al. 1991). The highest levels were found in kidney DNA where some adducts persist for more than 16 days, while adducts were repaired in liver and spleen after 5 days (Pfohl-Leszkowicz et al. 1993).

According to Zlender et al. (2009), OTA induced the formation of 8-hydroxy-2'-deoxyguanosine (8-OHdG), which is a commonly used biomarker of DNA oxidative damage. In Madine-Darby canine kidney (MDCK) cells, the effect of OTA on DNA was investigated by detecting the occurrence of DNA damage by alkaline single cell gel electrophoresis assay (comet assay). It was observed that OTA induced single-strand DNA breaks in a concentration-dependent manner (Lebrun and Follmann 2002). It has been reported that OTA causes large deletion mutations in the process of homologous recombination repair for double strand breaks (Hibi et al. 2013). OTA has also been reported to induce chromosomal mitotic recombination in *Drosophila* (Vogel and Nivard 1993).

Evidence against Direct Genotoxic Mechanism

The potential of OTA to form covalent DNA adducts has been subjected to debate, due to conflicting data in the literature. DNA adducts of OTA have not been identified by several researchers *in vitro* under physiological conditions neither by radiolabeling (Gautier et al. 2001a) nor by liquid chromatography-tandem mass spectrometry (LC-MS/MS) techniques (Mally et al. 2004; 2005). Moreover, DNA adducts were not detected in rats exposed to ^{14}C-labeled OTA using the highly sensitive accelerator mass spectrometry technique (Zepnik et al. 2003; Mally et al. 2004). In an investigation using ^{3}H-labeled OTA, no radioactivity was detected in DNA of rat kidney based upon liquid scintillation counting with a limit of detection of 2.7 adducts/109 DNA bases (^{32}P-postlabeling experiments showed lesions at levels 3-17-fold higher than this value) (Gautier et al. 2001a).

In the presence of cytochrome P450, OTA undergoes oxidative dechlorination to generate the electrophilic quinone OTQ that reacts covalently with glutathione (GSH) to generate the GSH conjugate (Dai et al. 2002). The quinone OTQ can undergo either two-electron reduction by the action of the NAD(P)H:quinone reductase to form hydroquinone OTHQ, or undergo one electron-reduction to yield a semiquinone radical, which in turn induce DNA breaks, lipid peroxidation (LPO) and exocyclic adducts.

OTA-induced superoxide radical ($O_2^{\bullet-}$) toxicity liberate ferrous ion (Fe^{2+}), or other reducing equivalents that cause reductive dehalogenation of OTA to generate the reactive aryl radical. Reaction of aryl radical with hydrogen donor further leads to the production of non-chlorinated OTB metabolite. OTB has been reported to cause abortive mitosis, formation of multiple nucleated cells, and cell degeneration *in vitro* in epithelial cells of monkey kidney (Steyn et al. 1975). Hence, the recent data indicate that OTA causes genotoxicity through an indirect mechanism that involves oxidative DNA damage.

(ii) Epigenetic or Non-Genotoxic Mechanism

The mechanism of OTA cytotoxicity mainly depends on its concentration and period of its exposure to the organism. At lower concentrations, OTA affects the cell in a specific way through interactions with secondary messengers and other cell targets, which subsequently activate mitogen-activated protein kinases (MAPKs) (Schramek et al. 1997) and disturb cellular pH homeostasis (Gekle and Silbernagl 1996). At relatively high OTA concentrations, severe disturbances in the cell metabolism has been observed, such as stimulation of lipid peroxidation by complexing and facilitating reduction of iron (Omar et al. 1996); inhibition of phenylalanine metabolizing enzymes, thereby affecting DNA, RNA, protein synthesis (Creppy et al. 1984) and inhibiting mitochondrial ATP production by acting as a competitive inhibitor of carrier proteins located in the inner mitochondrial membrane (Wei et al. 1985).

Lipid Peroxidation

OTA has been found to induce oxidative damage *in vitro* (Schaaf et al. 2002) and *in vivo* (Petrik et al. 2003). Lipid peroxidation is the first effect of oxidative damage to the cell. OTA stimulates lipid peroxidation by forming ferric ion (Fe^{3+})-OTA complex and leaking calcium ions (Ca^{2+}) from Ca^{2+}-loaded microsomes. Mobilization of Fe^{3+} and Ca^{2+} by OTA lead to uncoupling of oxidative phosphorylation and generation of toxic hydroxyl ($OH^{\bullet}$) free radical in the presence of NADPH-cytochrome P450 reductase system. OTA-induced reactive oxygen species (ROS) production is preceded by loss of cell viability and initiation of DNA damage (Hasinoff et al. 1990).

Protein Synthesis Inhibition

OTA has been shown to inhibit protein synthesis in all experimental systems tested including bacteria, yeast cells, mammalian cells *in vitro* and animals *in vivo* (Dirheimer and Creppy 1991). The inhibition of protein synthesis is likely to be involved in most of the acute toxic effects of OTA (Dirheimer and Creppy 1991; WHO, 2001; Schilter et al. 2005). At lower doses, the effect of OTA on protein synthesis inhibition is less obvious. It has been observed that at low concentrations, OTA does not affect protein synthesis (Monnet-Tschudi et al. 1997; Gekle et al. 2005).

Genes involved in the negative regulation of protein synthesis, i.e., negative regulator of protein synthesis, the prostaglandin F2 receptor (Orlicky 1996) and eukaryotic factor for translocation onset Eif4ebp1 (Gingras et al. 1999), are over-expressed during OTA intoxication. On the contrary, the elongation factor eEF1A-1, gene involved in promoting the GTP dependent binding of aminoacyl-tRNA at the ribosomal site A, during protein synthesis is under-expressed (Petroulakis and Wang 2002).

Mitochondrial Dysfunctions

Mitochondria are the most susceptible organelle to OTA toxicity. OTA alters the mitochondrial transmembrane potential which leads to the release of cytochrome c and induction of cell apoptosis (Assaf et al. 2004). OTA competitively inhibits ATPase, succinate dehydrogenase and cytochrome c oxidase (Rumora and Grubisic 2009). Phosphoenolpyruvate carboxykinase is also affected acutely by OTA toxicity. The mechanism appears to be *via* specific degradation of mRNA encoding for these enzymes (Dirheimer and Creppy 1991).

Mitochondrial dysfunctions exhibited by OTA are mainly responsible for its nephrotoxic effect. Wei et al. (1985) conducted *in vitro* studies by applying direct OTA treatments on mitochondrial preparations and showed that OTA affects respiration and oxidative phosphorylation through an impairment of the mitochondrial membrane and an inhibition of succinate. Aleo et al. (1991) treated isolated renal proximal tubules with large concentrations of OTA and observed an inhibition in mitochondrial respiration.

Protein Damage

OTA induced oxidative stress causes carbonylation of the side amino acid groups of proteins, forming protein carbonyls that are widely used as markers of protein oxidation (Davies 2003). Due to the abundance of protein carbonyl content in mammalian cells, cytoskeletal proteins like actin are common targets for a variety of ROS (Dalle-Donne et al. 2006).

Alteration of Signal Transduction

- *Calcium homeostasis:* Cytosolic calcium (Ca^{2+}) is maintained at low concentrations through active sequestration in extracellular medium, mitochondria and endoplasmic reticulum. Disruption of Ca^{2+} homeostasis leads to its sustained increase in cytosol that can cause cell injury by activating proteases and phospholipases. In addition, Ca^{2+} is known to be one of the major players in cell signaling, e.g., through the activation of regulatory pathways involving calmodulin and protein kinase C. Benesic et al. (2000) observed that in immortalized human kidney epithilial cells-1 (IHKE-1), nanomolar concentrations of OTA induced rapid and reversible oscillations in intracellular calcium. In the same system, OTA potentiated the calcium-dependent stimulation of cell proliferation induced by epidermal growth factor and angiotensin II suggesting that OTA may interfere with calcium-dependent cell signaling leading to increased cell proliferation.
- *Mitogen-activated protein kinases (MAPKs):* The activation of these proteins is related to cell-cycle arrest. The actual molecular mechanism responsible for OTA-mediated activation of MAPK cascades has not yet been elucidated. However, it seems to involve oxidative stress. ROS or redox disturbances stimulate redox-sensitive signaling molecules such as MAPKs and modify expression of various genes (Rumora and Grubisic 2009). In renal cells, OTA is able to stimulate MAPK activities promoting opposite effects, either proliferation or apoptosis. The balance between different MAPKs will determine the fate of the exposed cells (Schilter et al. 2005).
- *Apoptosis:* Tissue damage characterized by cell death is the common histopathological feature of OTA toxicity. OTA induced cell death has been identified in the form of both necrosis and apoptosis. The dosage of the toxin mainly determines whether cell die by necrosis or apoptosis. At lower dose, OTA induces cell death mainly through apoptosis, whereas at higher dose, stimulation of both necrosis and apoptosis has been reported (Sauvant et al. 2005a; Ozcan et al. 2014).

Apoptosis refers to programmed cell death, an active process to be differentiated from necrosis. Induction of apoptosis is considered to be one of the major cellular mechanisms behind OTA-induced multiple toxic effects, especially in its main target, i.e., kidney. Several studies have suggested that

OTA induces apoptosis in animal tissues *in vivo* and in cell cultures (WHO 2001; O'Brien and Dietrich 2005; Schilter et al. 2005).

OTA mediated apoptosis involve chromatin condensation, DNA degradation, membrane blebs formation and cell collapse. The well-known pathways of apoptosis include the extrinsic pathway mediated through death receptors, the intrinsic pathway centered on mitochondria, and the endoplasmic reticulum (ER)-stress pathway (Figure 2).

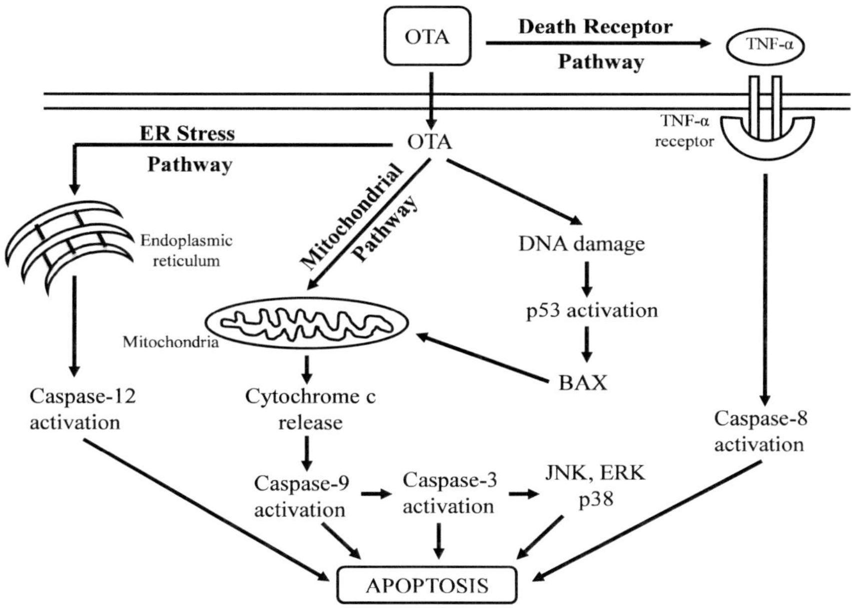

Figure 2. Pathways involved in OTA-mediated apoptosis.

The extrinsic pathway or death receptor pathway involves binding of the ligands to death receptors at the plasma membrane, the activation of caspase-8 and further activation of downstream caspases to induce apoptosis. The main death receptors include Fas and TNF-α receptors (TNFR). The intrinsic or mitochondrial pathway is the major apoptotic pathway in OTA toxicity. The cellular stress activates the proapoptotic members of Bcl-2 family proteins, such as BAX and BAK, which form pores in the outer mitochondrial membrane, resulting in the release of cytochrome c into the cytosol. The cytochrome c associates with apoptotic protease activating factor-1 (Apaf-1) and procaspase-9 to form the apoptosome, which further activates caspase-3 and various steps of programmed cell death (Schwerdt et al. 1999; Pinton et al.

2008). OTA toxicity may also involve the ER-stress pathway. The initiator caspase in the ER pathway is caspase-12, which is localized at the cytosolic face of the ER and is activated by ER stress.

OTA stimulates physiological pro-apoptotic stimuli such as TNF-α release (Al-Anati et al. 2010) and reduce anti-apoptotic signaling cascades (Assaf et al. 2004). Gekle et al. (2000) demonstrated that OTA interfere with physiological pro-apoptotic stimuli such as TNF-α by enhancing its effect on c-Jun-N-terminal kinase (JNK) activity in a concentration-dependent manner in MDCK-C7 cells and potentiate the effect of TNF-α on caspase-3-activity. In a recent study, Liang et al. (2015) suggested that a MAPK family member, i.e., apoptosis signal-regulating kinase 1 (ASK1) plays an important role in OTA-mediated oxidative stress as well as ER-stress induced cell apoptosis. OTA induced oxidative stress triggers ASK1 activation. ASK1 is also found capable of amplifying ROS generation, probably due to the generation of oxidants, JNK activation and the decreased expression of transcription factors of certain antioxidants.

Another important molecular mechanism responsible for the toxicity of OTA involves p53 gene. p53 is the best known tumor suppressor protein that controls cell cycle checkpoints, apoptosis and DNA repair. This protein is also involved in cellular pro-survival response (Amin et al. 2010). The genotoxic stress induced by OTA triggers p53-dependent apoptotic process in the cells. Activation of this transcription factor leads to cell cycle arrest, allowing DNA repair, or apoptosis in case of severe DNA damage (Finlan and Hupp 2005). p53 also activates apoptotic response genes, such as BAX, PUMA or NOXA, whose products trigger the mitochondrial pathway of apoptosis (Bouaziz et al. 2008). Li et al. (2011) observed that OTA induces p53 activation and apoptosis in human and monkey kidney cells. It was further demonstrated that the pro-survival function of p53 was associated with suppression of JNK activation which was found to play a critical role in OTA-induced apoptosis.

Necrosis

Necrosis has been considered as an unplanned and uncontrolled type of cell death. The serine/threonine kinase receptor-interacting protein 1 (RIP 1) plays a major role in necrosis initiation. In addition, Ca^{2+} and ROS are the main players in necrotic cell death. Increased cytosolic Ca^{2+} levels due to OTA toxicity lead to mitochondrial Ca^{2+} overload and activation of proteases and phospholipases. ROS generated due to OTA toxicity initiate damage to cell macromolecules, mitochondrial dysfunction, ion balance deregulation and loss of membrane integrity. Finally, necrotic cells release different factors that lead to recognition and engulfment by phagocytes and the subsequent immunological response (Vanlangenakker et al. 2008). Figure 3 represents the various pathways involved in OTA-mediated cell necrosis.

Carcinogenicity

Tumors induced by OTA are found to be highly metastatic. Mitogen-activated protein kinases (MAPKs) are involved in directing cellular responses to a diverse array of stimuli and regulation of cell functions including proliferation, differentiation, gene expression, cell survival and apoptosis. The most important members of MAPK family include c-Jun-N-terminal kinase (JNK), p38 kinase and extracellularly-regulated protein kinase (ERK). Using *in vitro* systems, OTA was found to induce phosphorylation of ERK, JNK and p38 (Gekle et al. 2005; Sauvant et al. 2005a). The isoforms of ERK1/2 have been implicated in cancer development including renal carcinoma (Oka et al. 1995). In MDCK-C7 cells from collecting duct and renal proximal tubule cell lines, OTA was found to increase the phosphorylation and activity of ERK1/2. In the same cellular systems, other MAPKs, i.e., JNK and p38 kinase, were also activated by OTA treatments. Unlike ERK1/2, JNK and p38 kinase do not seem to act as promitotic signals, but are predominantly implicated in stress response and apoptosis (Schilter et al. 2005).

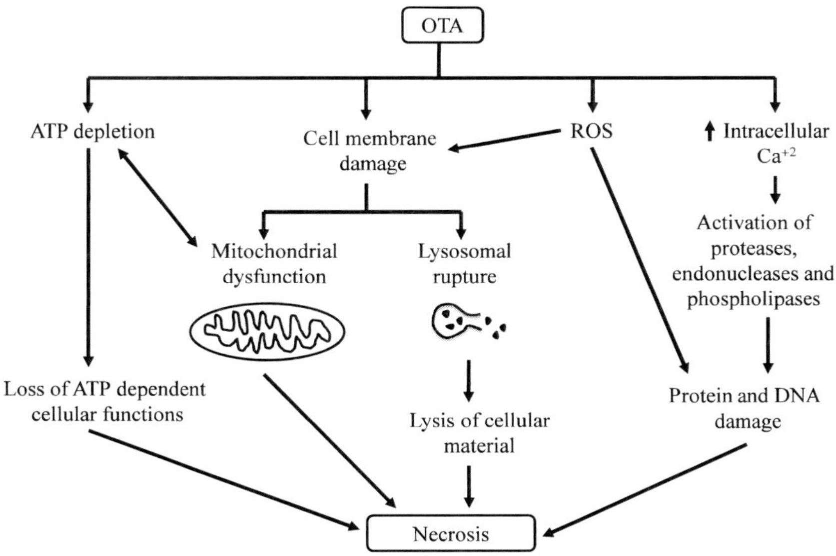

Figure 3. Pathways involved in OTA-mediated cell necrosis.

It has also been observed that OTA increases the expression of insulin-like growth factor-1 (IGF-1) receptor and phosphorylates phosphoinositide dependent kinase-1 (PDK-1) and protein kinase C (PKC). A direct association has been observed between cell proliferation, cell survival, anti-apoptotic activity and renal cancer development with the stimulation of the IGF-1 system (Sridhar et al. 2005; Takahashi et al. 2005), an increased PKC activity (Kaelin 2003; Kim and Kaelin 2003) and the activation of MAPKs (Oka et al. 1995; Dong et al. 2004).

In kidney cells of rodents, it has been shown that mTOR/AKT pathways are significantly deregulated after exposure to OTA, possibly contributing to carcinogenicity (Stemmer et al. 2009). OTA increases the activity of transcription factor, i.e., NF-κB which is involved in the transcription of genes responsible for cell growth and inhibition of apoptosis. Sustained activation of NF-κB has been associated with cancer development (Klaunig and Kamendulis 2004). In Figure 4, the information pertaining to the OTA mediated carcinogenicity involving a complex network of interacting epigenetic mechanisms has been integrated. The detailed explanation regarding OTA mediated carcinogenicity in various organs has been discussed later.

According to a study conducted by Czakai et al. (2011), OTA promotes genomic instability and tumorigenesis through interference with cell division. OTA interferes with cell division, leading to cell death at mitosis or aberrant exit from mitosis without nuclear division. The disruption of mitosis by OTA is accompanied by aberrant chromosome condensation and segregation and altered post translational modification of histone and non-histone proteins. Histone acetyltransferases (HATs) are the primary cellular targets of OTA. HAT complexes participate in a number of key nuclear processes including chromosome decondensation, DNA replication, DNA damage repair, and gene-specific transcription regulation. HATs also act as tumor suppressors. Mitotic delay in response to OTA occurs as a consequence of altered histone acetylation patterns and associated hyper-condensation of mitotic chromosomes.

OTA acts outside mitosis (e.g., in G_1 or S phase) by altering transcription or post translational acetylation of mitotic regulators rather than by directly altering mitotic chromosome structure. Post translational acetylation of non-histone proteins through OTA-mediated inhibition of HATs is responsible for multiple aspects of OTA toxicity from mitotic disruption to altered transcription and stress/DNA damage response. Figure 5 explains the events involved in renal tumor formation by OTA.

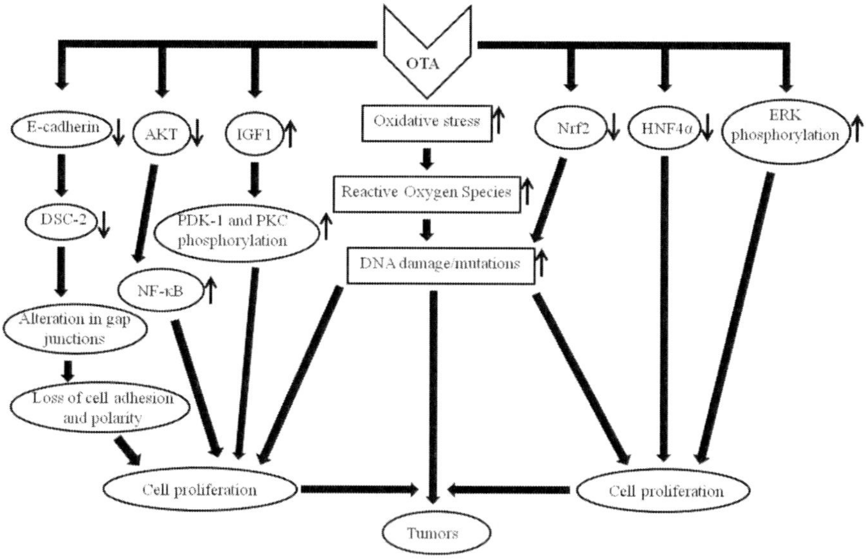

Figure 4. Proposed model for OTA-mediated carcinogenecity involving a complex network of interacting epigenetic mechanisms. OTA effects are shown as: decrease ↑ and increase ↓ .

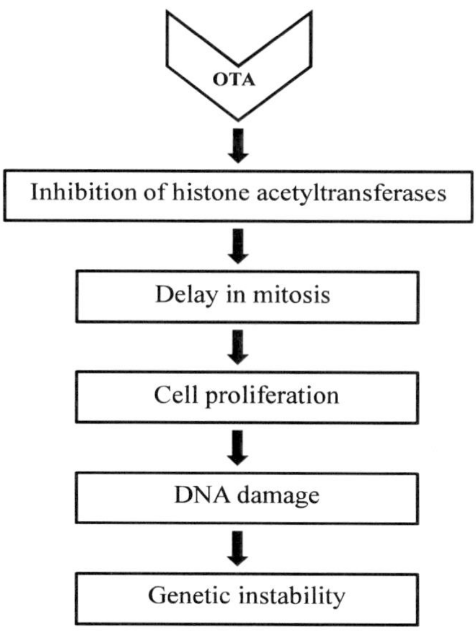

Figure 5. Sequence of proposed key events in renal tumor formation by OTA.

Cyclooxygenase-1 (COX-1) plays a crucial role in the preservation of renal function and is constitutively expressed in all tissues. Cyclooxygenase-2 (COX-2) is less constitutively expressed in all tissues but can readily be induced (Prescott and Fitzpatrick 2000). Its induction has been related to tumor cell growth and progressive cell injury (Zha et al. 2004). OTA decreases protection of kidney function due to inhibition of COX-1, and increases risk of cancer due to over expression of COX-2 (Pfohl-Leszkowicz and Manderville 2007).

Teratogenicity

OTA is a potent teratogen in mice, rats, hamsters, chickens and rabbit. Maternal OTA cross the placenta and cause abnormalities in central nervous system, eyes and skeleton of developing foetus (Kuiper-Goodman and Scott 1989). The teratogenic effect of OTA is dependent upon the dose as well as the period of gestation during which the toxin administered. It has been observed that gestation days 6 and 7 are the most critical for the induction of OTA-mediated teratogenicity in rats (Patil et al. 2006). The effects are more severe at the time of neurulation and have not been observed after closing of neuropores. This might be due to specific action of the toxin on the developing nervous system or reduced placental transfer of the toxin later in development (O'Brien et al. 2005).

OTA produces various gross, skeletal and visceral anomalies particularly at the higher dose level. The gross anomalies observed in rabbit include wrist drop, rudimentary tail, knuckling of fetlock and agenesis of tail. Skeletal anomalies included agenesis of caudal vertebrae, incomplete ossification of skull bones and wavy ribs. The soft tissue anomalies included internal hydrocephalus, microphthalmia and kidney agenesis (Wangikar et al. 2004).

For the teratogenic mechanism of OTA, role of maternal protein deprivation, impaired glycolysis and inactivated phosporylase-b-kinase have been suggested by researchers (Mayura et al. 1983). According to Hood et al. (1976), OTA inhibits mitochondrial respiration and has a direct effect on foetus, rather than having its action mediated through an effect on the dam. OTA mediated teratogenicity is the result of inhibition of DNA, RNA and protein synthesis (Marquardt and Frohlich 1992), intracellular transport and subsequent lipid peroxidation resulting from generation of free radicals (Wei and Sulik 1993), direct effect on osteoblasts and osteoclasts and interference in the calcium homeostasis (Khan et al. 1989) involved in excessive embryonic cell death.

The toxic effects of OTA on various organs and their possible mechanisms of action are discussed below.

Nephrotoxicity

OTA is nephrotoxic to birds and mammals (Krogh 1980; Sujani and Sriraman 2000), but not to adult ruminants (Ribelin et al. 1978). At higher doses, OTA affects both renal morphology and function, as indicated by increased weight, urine volume, blood urea nitrogen, urinary glucose, and proteinuria (Hatey and Galtier 1977). Elling et al. (1985) observed that OTA results in condensation of cellular material with disappearance of membranes and continuous desquamation in the lower part of the proximal convoluted tubules (PCTs).

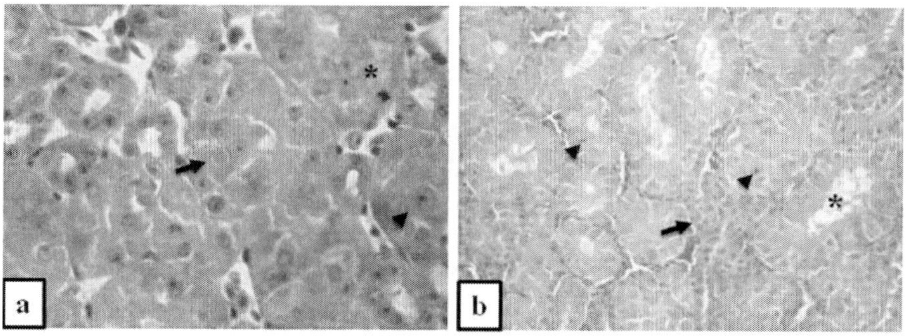

Figure 6. Photomicrograph showing effect of OTA on kidney of Japanese quail. (a) Kidney section showing PCTs swelling, increased cytoplasmic granularity (star), presence of hyaline body in the epithelial cell (arrow) and karyomegaly (arrowhead). *H & E x 330.* (b) Kidney section showing PCTs swelling (star), mononuclear cells infiltration (arrow) and mitotic bodies in the epithelial cells (arrowhead). *H & E x 132.*

In target kidney cells, peroxisomes lose membrane integrity and leak their contents into the cytosol. The toxic effects of OTA are more pronounced in proximal convoluted tubules than distal convoluted tubules. Patial et al. (2013) observed predominant degenerative changes such as epithelial cell swelling in PCTs due to increase in cytoplasmic granularity and varying degree of cytoplasmic vacuolations along with marked karyomegaly (Figure 6a, b). Mitochondria were primarily affected organelle in PCT cells, with marked swelling, pleomorphism and reduction in their number due to OTA toxicity (Figure 7a, b). OTA also caused degranulation of rough endoplasmic reticulum, thickening of glomerular basement membrane and fusion of podocyte foot processes (Figure 7c, d). Nuclear changes involved indistinct nuclear membrane with indentation or blebbing, uneven distribution of chromatin material along with margination and karyomegaly (Figure 7e, f).

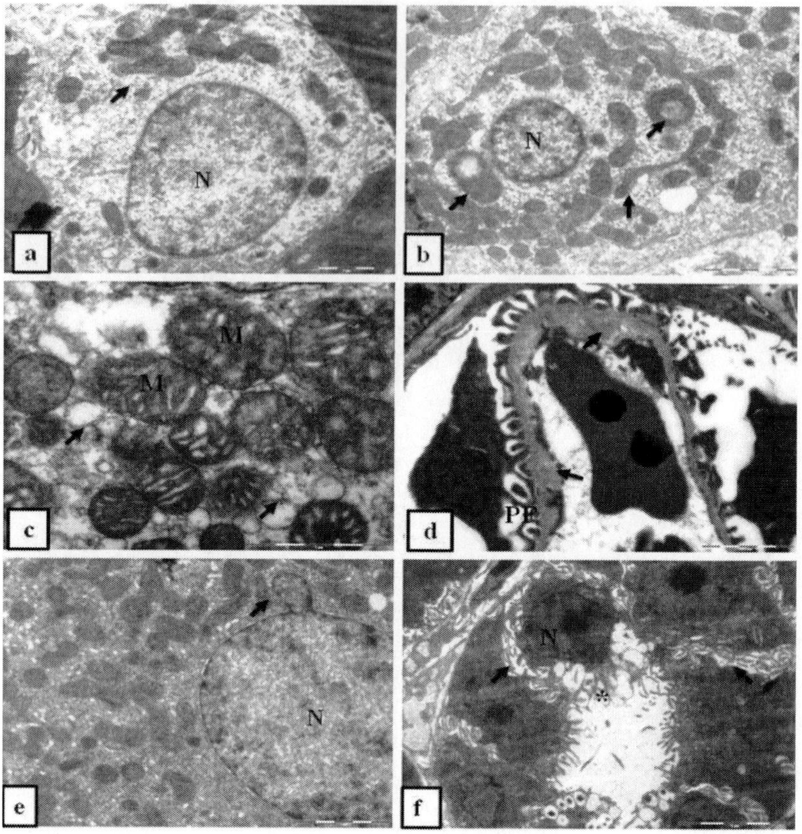

Figure 7. Transmission electron microscopy photographs depicting ultrastructural alterations caused by OTA in the kidney of Japanese quail. (a) Kidney section showing a marked decrease in the number of mitochondria (arrow) along with disappearance of other cell organelles from the cytoplasm of a PCT epithelial cell. (b) Kidney showing degenerating nucleus (N) and mitochondrial pleomorphism with ring shape and elongated mitochondria (arrows). (c) Kidney section showing degenerating mitochondria with formation of vesicle like structures (M) in the mitochondrial matrix and severe dilation and degranulation of RER (arrows). (d) Kidney showing uneven thickness of glomerular basement membrane (arrows) and fusion of podocytic foot processes (PF). (e) Kidney section showing bleb formation of nuclear membrane (arrow). (f) Kidney showing epithelial cells with pyknotic nuclei and indented nuclear membranes (N), widening of intercellular junctions (arrows) and loss of microvilli in the lumen of PCTs (star).

OTA severely affects epithelial brush border of PCTs with blunting and variable loss of microvilli. This leads to the accumulation of debris in the tubular lumen and narrowing of tubular space due to swelling of microvilli (Figure 7f). In broiler chicks, exposure of 0.5 mg OTA per week for 4 weeks induced glomerulonephrosis, tubulonephrosis, focal tubular epithilial cell proliferation and multiple adenoma like structures in the renal parenchyma (Biro et al. 2002). OTA has also been shown to induce karyomegaly in the proximal tubular cells of the S3 segment. These lesions may result from the promotion of cell-cycle progression and activation of the G2/M checkpoint in response to DNA damage (Rached et al. 2007; Hibi et al. 2011).

OTA induced nephropathy occurs in a progressive manner by tubular degeneration, interstitial fibrosis, and impaired renal function. The post proximal parts of nephron are mainly affected by the acute OTA exposure, whereas prolonged exposure mostly damages the proximal tubules (Gekle and Silbernagl 1996). OTA induced chronic interstitial nephropathy involves phenomena like loss of cells and epithelial tightness, inflammation, fibrosis, epithelial-to-mesenchymal transition and cell death in proximal tubules (Sauvant et al. 2005b). OTA also causes defect in the organic anion transport mechanism located on the brush border membranes of the proximal convoluted tubules and basolateral membranes (Gupta 2012).

Mitochondrial dysfunction has been shown to be involved in the development of OTA-induced toxicity in PCT cells (Aleo et al. 1991). OTA inhibits mitochondrial oxidative phosphorylation by acting as a competitive inhibitor of carrier proteins located in the inner mitochondrial membrane (Solcan et al. 2008). The inibition of protein synthesis and the damaged energy production in the mitochondria could be considered as the most important factors for degenerative changes in the epithelial cells of proximal tubules. Oxidative damage including lipid peroxidation also resulted in OTA mediated nephrotoxicity (Dirheimer and Creppy 1991; Patial et al. 2013).

In kidneys, genes encoding the antioxidative enzymes are not fully expressed during OTA toxicity. These genes contain antioxidant regulatory element (ARE) in their promoter region. The ARE is recognised by the transcription factor, i.e., nuclear factor-erythroid 2 p45-related factor 2 (Nrf 2) which is involved in the expression and induction of antioxidant genes. OTA alter the expression of genes regulated by the transcription factor Nrf 2 (Figure 4), thereby affecting the cell's antioxidative defense which makes the cell more sensitive to oxidation (Kensler et al. 2007). There are several possible mechanisms of OTA-induced Nrf 2 inhibition, i.e., inhibition of Nrf 2 nuclear translocation; inhibition of Nrf 2 DNA binding; or epigenetic effects preventing normal Nrf 2-dependent transcription. Inhibition of Nrf 2 by OTA is the mechanism for both its nephrotoxicity and carcinogenicity (Limonciel and Jennings 2014).

OTA induces collagen secretion in damaged epithelial cells of human proximal tubules, thus impairing the function of kidneys. OTA also reduces the glomerular filtration rate of kidneys. OTA affects carbohydrate metabolism, particularly gluconeogenesis in kidneys by reducing the renal mRNA coding for phosphoenolpyruvate carboxykinase. The link between citric acid cycle and gluconeogenesis, i.e., phosphoenolpyruvate carboxykinase makes possible the transformation of citric acid cycle intermediates and their precursors into glucose and glycogen. Interference with this rate-limiting step in gluconeogenesis plays a key role in the development of functional damage to the kidney (Ueno 1991).

Hepatotoxicity

Main pathological changes in the liver of rats (Aydin et al. 2003) and Japanese quail (Patial 2011) exposed to OTA prominently include granular or vacuolated degeneration and necrosis of the hepatocytes, sinusoidal and central vein dilation, bile duct proliferation, enlargement of periportal areas with mononuclear cell inflammatory infiltration and mild degree proliferation of fibrous tissue (Figure 8a, b).

Atroshi et al. (2000b) observed that OTA results in increased activities of aspartate aminotransferase, alanine aminotransferase and alkaline phosphatase which are used as sensitive markers of hepatocellular necrosis. The occurence of OTA induced hepatocellular carcinoma in mice has also been reported (Bendele et al. 1985). Abdu (2011) observed that treatment of rats with OTA alter the function of the hepatocytes as indicated by the increased serum levels of total bilirubin and alanine transaminase activity. The elevation of the serum level of bilirubin is also an indicator for the disturbance of the hepatic cellular function (Green and Flamm 2002).

According to Abdel-Wahhab et al. (2005), OTA induces oxidative stress in the hepatic cells. The cellular biomolecules *viz.*, lipids, proteins, nucleic acids and carbohydrates are known to undergo oxidative modifications (Zamzami et al. 1995). OTA affects the cellular viability by inhibiting the synthesis of cellular proteins and phospholipids (Atroshi et al. 2000a). Robbins and Angell (1976) reported that swelling and vacuolization of the hepatocytes due to OTA toxicity may reflect failure of membrane ion pump due to ATP deficiency allowing these cells to accumulate fluids.

OTA hepatotoxicity is correlated to the deregulation of important pathways like primary bile acid biosynthesis and metabolism of xenobiotics by cytochrome P450. Marin-Kuan et al. (2006) observed a significant reduction in the expression of genes regulated by the transcription factor, i.e., hepatocyte nuclear factor (HNF4α) (Figure 4). HNF4α-regulated genes encode for proteins of various functions including xenobiotic metabolizing enzymes and transport proteins. Their reduced expression result in a decreased detoxification and excretion of toxic xenobiotics. OTA also decreases the expression level of glutathione S-transferase, reported to be significantly down-regulated during the development of hepatocellular preneoplastic foci in rats (Qi et al. 2014).

OTA induced liver injury reduces the cell's ability to counter balance oxidative stress which leads to the reduction of E-cadherin and down regulation of desmocollin-2 (DSC-2) (Figure 4). This causes alteration in gap junctions and loss of cell adhesion and polarity, thereby triggering the promotion of liver tumors (Gagliano et al. 2006).

Immunotoxicity

OTA causes reduction in the size as well as alterations in the function of immune organs, such as thymus, spleen, bursa and lymph nodes (Figure 8c, d). OTA increases the secretion of pro-apoptotic cytokines, i.e., tumor necrosis factor-α (TNF-α) and interleukin-6 (IL-6) which trigger organ size reduction and cell depletion.

Monocyte/macrophages play a major role in fibrin accumulation associated with immune-inflammatory processes through the production of tissue factor (TF) and plasminogen activator inhibitor-2 (PAI-2). Bacterial lipopolysaccharides (LPS) act as stimulating agent for monocytes/ macrophages. Rossiello et al. (2008) observed that OTA induces suppression of LPS-induced monocyte/macrophage TF and PAI-2 response and results in the impairment of blood clotting activation, fibrin formation and cell activation leading to a diminished inflammatory response and attenuated cell-mediated immune response.

OTA causes bone marrow depression and hypocellularity due to migration of white blood cells and immune cells to OTA damaged target organs. It also reduces the number of bone marrow macrophage-granulocyte progenitors and hematopoietic stem cells and decreases erythropoiesis (Boorman et al. 1984). Moreover, OTA and its metabolites depress immune cell metabolic activity, proliferation, differentiation and membrane integrity (Muller et al. 2003).

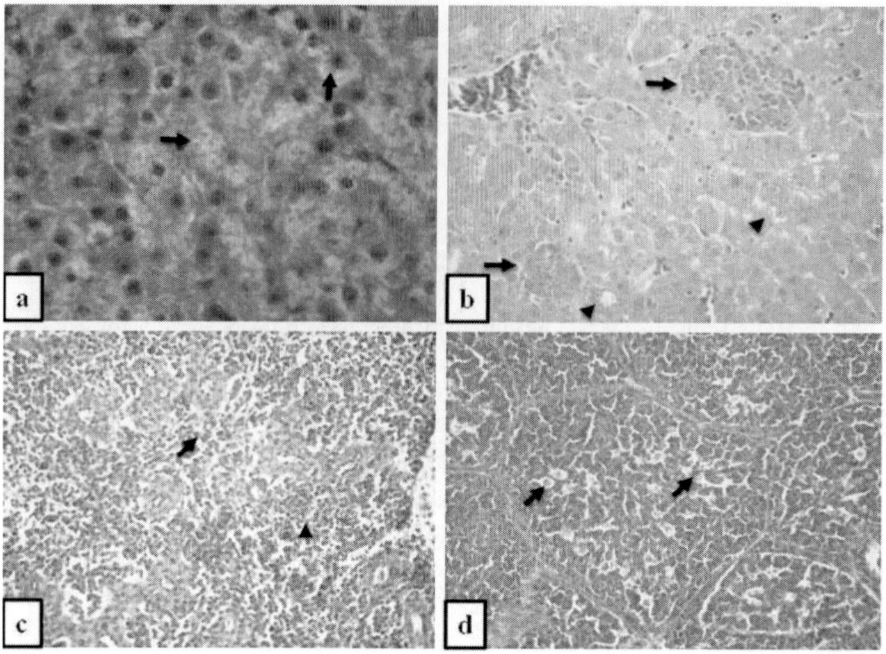

Figure 8. Photomicrograph showing effect of OTA on different organs of Japanese quail. (a) Liver section showing swollen hepatocytes (arrow) with cytoplasmic granularity. *H & E x 330.* (b) Liver section showing hepatocytic vacuolation (arrowhead) and necrosis with the infiltration of mononuclear cells and heterophils (arrows). *H & E x 132.* (c) Spleen section showing marked depletion of white pulp (arrow) along with RE cell hyperplasia (arrowhead). *H & E x 132.* (d) Section of bursa showing depletion of lymphoid element in the medullary region (arrows). *H & E x 132.*

Assaf et al. (2004) showed that OTA could induce immunosuppression *via* marked apoptosis in human lymphocytes. Liu et al. (2012) hypothesized that OTA exposure in blood might induce cell cycle arrest and apoptosis in human peripheral blood mononuclear cells (hPBMC). The authors reported that in hPBMC, oxidative stress is involved in OTA-induced human immunotoxicity and concluded that OTA-induced DNA damage and G1 arrest play a major role in OTA-mediated carcinogenicity.

Neurodegeneration

OTA is also responsible for the pathogenesis of neurodegenerative diseases such as Alzheimer's and Parkinson's disease. OTA induces its deleterious effect on brain by affecting proliferation and migration of neurons and reducing DNA content in the developing brain. Neurotoxicity has been reported to be more pronounced in the ventral mesencephalon, hippocampus, and striatum than in the cerebellum (Doi and Uetsuka 2011). The mechanisms of toxicity include inhibition of protein synthesis, mitochondrial impairment, oxidative stress, bioenergetic compromise, DNA damage and apoptosis (Dirheimer and Creppy 1991; Gautier et al. 2001b; Sava et al. 2006). OTA leads to the activation of N-methyl-D-aspartate receptor (NMDAR) which is a glutamate receptor and an ion channel protein found in nerve cells. Persistent activation of NMDAR by OTA results in excitotoxicity mediated by the neurotransmitter glutamate in brain accounting for the vulnerability of striatum and pallidum (Turski and Turski 1993). In addition, Ca^{2+} enters inside the neurons through NMDAR and lead to the production of free radicals in mitochondria resulting in its depolarization (Greenamyre et al. 1999).

CONCLUSION

The information reviewed above suggests that OTA is unlikely to act through a single, well defined mechanism of action. Oxidative stress may constitute a mechanism for many of the effects associated with OTA treatment. Oxygen radicals may either originate from reactions directly involving OTA (with Fe^{3+}) or may be the consequence of the OTA-mediated reduction of antioxidant cellular defenses, which could no longer cope with radicals generated over the normal oxidative metabolism. The production of oxygen radicals is likely to produce macromolecular damage including lipid peroxidation and DNA damage. This latter effect may result in mutation and cancer initiation. In addition, ROS modify gene expression through various mechanisms involving mobilization of MAPK cascade and activation of transcription factors such as NFκB. Such biological effects may either lead to apoptosis or to cancer development.

REFERENCES

Abdel-Wahhab, M. A.; Abdel-Galil, M. M. & El-Lithey, M. (2005). Melatonin counteracts oxidative stress in rats fed an ochratoxin A contaminated diet. *Journal of Pineal Research,* 38, 130-135.

Abdu, S. B. (2011). The protective role of ajwa date against the hepatotoxicity induced by ochratoxin A. *Egyptian Journal of Natural Toxins,* 8(1, 2): 1-15.

Al-Anati, L.; Essid, E.; Stenius, U.; Beuerlein, K; Schuh, K & Petzinger, E. (2010). Differential cell sensitivity between OTA and LPS upon releasing TNF-α. *Toxins,* 2(6), 1279-1299.

Aleo, M. D.; Wyatt, R. D. & Schnellman, R. G. (1991). Mitochondrial dysfunction is an early event in ochratoxin A but not oosporein toxicity to rats' renal proximal tubules. *Toxicology and Applied Pharmacology,* 107, 73-80.

Amin, A. R.; Thakur, V. S.; Gupta, K.; Jackson, M. W.; Harada, H.; Agarwal, M. K.; Shin, D. M.; Wald, D. N. & Agarwal, M. L. (2010). Restoration of p53 functions protects cells from concanavalin A-induced apoptosis. *Molecular Cancer Therapeutics,* 9(2), 471-479.

Assaf, H.; Azouri, H. & Pallardy, M. (2004). Ochratoxin A induces apoptosis in human lymphocytes through down regulation of Bcl-xL. *Toxicological Sciences,* 79, 335-344.

Atroshi, F.; Biese, I.; Saloniemi, H.; Ali-Vehmas, T.; Saari, S.; Rizzo, A. & Veijalainen, P. (2000a). Significance of apoptosis and its relationship to antioxidants after ochratoxin A administration in mice. *Journal of Pharmacy & Pharmaceutical Sciences,* 3, 281-291.

Atroshi, F.; Rizzo, A.; Sankari, S.; Biese, I.; Westermarck, T. & Veijalainen, P. (2000b). Liver and enzyme activities of rats exposed to ochratoxin A and T-2 toxin with antioxidants. *Bulletin of Environmental Contamination and Toxicology,* 64, 586.

Aydin, G.; Ozcelik, N.; Cicek, E. & Soyoz, M. (2003). Histopathological changes in liver and renal tissues induced by ochratoxin A and melatonin in rats. *Human & Experimental Toxicology,* 22(7), 383-391.

Bendele, A. M.; Carlton, W. W.; Krogh, E. B. & Lillehoj, E. B. (1985). Ochratoxin A carcinogenesis in the (C57BL/&JXC*H)F_1 mouse. *Journal of the National Cancer Institute,* 75(4): 733.

Benesic, A.; Mildenberger, S. & Gekle, M. (2000). Nephritogenic ochratoxin A interferes with hormonal signalling in immortalized human kidney epithelial cells. *Pflugers Archiv,* 439(3), 278-287.

Biro, K.; Solti, L.; Barna-Vetro, I.; Bago, G.; Glavits, R.; Szabo, E. & Fink-Gremmels, J. (2002). Tissue distribution of ochratoxin A as determined by HPLC and ELISA and histopathological effects in chickens. *Avian Pathology,* 31, 141-148.

Boorman, G. A.; Hong, H. L.; Dieter, M. P.; Hayes, H. T.; Pohland, A. E.; Stack, M. & Luster, M. I. (1984). Myelotoxicity and macrophage alteration in mice exposed to ochratoxin A. *Toxicology and Applied Pharmacology,* 72, 304-312.

Bouaziz, C.; Sharafeldein, O.; El Golli, E.; Abid-Essefi, S.; Brenner, C.; Lemaire, C. & Bacha, H. (2008). Different apoptotic pathways induced by zearalenone, T-2 toxin and ochratoxin A in human hepatoma cells. *Toxicology,* 254: 19-28.

Castegnaro, M.; Canadas, D.; Vrabcheva, T.; Petkova-Bocharova, T.; Chernozemsky, I. N. & Pfohl-Leszkowicz, A. (2006). Balkan endemic nephropathy: role of ochratoxin A through biomarkers. *Molecular Nutrition & Food Research,* 50, 519-529.

Castegnaro, M.; Mohr, U.; Pfohl-Leszkowicz, A.; Esteve, J.; Steinmann, J.; Tillmann, T.; Michelon, J. & Bartsch, H. (1998). Sex- and strain-specific induction of renal tumors by ochratoxin A in rats correlates with DNA adduction. *International Journal of Cancer,* 77, 70-75.

Clark, H. A. & Snedeker, S. M. (2006). Ochratoxin A: its cancer risk and potential for exposure. Journal of Toxicology and Environmental Health, *Part B, Critical Reviews,* 9, 265-296.

Creppy, E. E.; Roschenthaler, R. & Dirheimer, G. (1984). Inhibition of protein synthesis in mice by ochratoxin A and its prevention by phenylalanine. *Food and Chemical Toxicology,* 22, 883-886.

Czakai, K.; Muller, K.; Mosesso, P.; Pepe, G.; Schulze, M.; Gohla, A.; Patnaik, D.; Dekant, W.; Higgins, J. M. G. & Mally, A. (2011). Perturbation of mitosis through inhibition of histone acetyltransferases: The key to ochratoxin A toxicity and carcinogenecity? *Toxicological Sciences,* 122(2), 317-329.

Dai, J.; Park, G.; Wright, M. W.; Adams, M.; Akman, S. A. & Manderville, R. A. (2002). Detection and characterization of a glutathione conjugate of ochratoxin A. *Chemical Research in Toxicology,* 15, 1581-1588.

Dalle-Donne, I.; Rossi, R.; Colombo, R.; Giustarini, D. & Milzani, A. (2006). Biomarkers of oxidative damage in human disease. *Clinical Chemistry,* 52(4), 601-623.

Davies, M. J. (2003). Singlet oxygen-mediated damage to proteins and its consequences. *Biochemical and Biophysical Research Communications,* 305, 761-770.

Dirheimer, G. & Creppy, E. E. (1991). Mechanism of action of ochratoxin A. *IARC Scientific Publications,* 115, 171-186.

Doi, K. & Uetsuka, K. (2011). Mechanisms of mycotoxin-induced neurotoxicity through oxidative stress-associated pathways. *International Journal of Molecular Sciences,* 12, 5213-5237.

Dong, J.; Everitt, J. I.; Lau, S. S. & Monks, T. J. (2004). Induction of ERK1/2 and histone H3 phosphorylation within the outer stripe of the outer medulla of the Eker rat by 2,3,5-tris-(glutathione-S-yl) hydroquinone. *Toxicological Sciences,* 80(2), 350-357.

Duarte, S. C.; Lino, C. M. & Pena, A. (2010). Mycotoxin food and feed regulation and the specific case of ochratoxin A: a review of the worldwide status. *Food Additives & Contaminants: Part A: Chemistry, Analysis, Control, Exposure & Risk Assessment,* 27, 1440-1450.

E. F. S. A. (European Food Safety Authority) (2006). Opinion of the scientific panel on contaminants in the food chain on a request from the Commission related to ochratoxin A in food, adopted on 4th April, 2006. *The EFSA Journal,* 365, 1-56.

Elling, F.; Nielsen, J. P.; Lillehoj, E. B.; Thomassen, M. S. & Stormer, F. (1985). Ochratoxin A induced porcine nephropathy: Enzyme and ultrastructure changes after short-term exposure. *Toxicon,* 23, 247-254.

Faucet, V.; Pfohl-Leskowicz, A.; Dai, J.; Castegnaro, M. & Manderville, A. (2004). Evidence for covalent DNA adduction by ochratoxin A following chronic exposure to rat and subacute exposure to pig. *Chemical Research in Toxicology,* 17, 1289-1296.

Finlan, L. E. & Hupp, T. R. (2005). The life cycle of p53: a key target in drug development. In: M., Los & S. B., Gibson (Eds.), *Apoptotic Pathways as Target for Novel Therapies in Cancer and Other Diseases.* Springer Academic Press, New York.

Follmann, W. & Lucas, S. (2003). Effects of ochratoxin A in a bacterial and a mammalian *in vitro* mutagenicity test system. *Archives in Toxicology,* 77, 298.

Gagliano, N; Donne, I. D.; Torri, C.; Migliori, M.; Grizzi, F.; Milzani, A.; Filippi, C.; Annoni, G.; Colombo, P.; Costa, F.; Ceva-Grimaldi, G.; Bertelli, A. A.; Giovannini, L. & Gioia, M. (2006). Early cytotoxic effects of ochratoxin A in rat liver: a morphological, biochemical and molecular study. *Toxicology,* 225(2-3): 214-224.

Galtier, P. (1977). Contribution of pharmacokinetic studies to mycotoxicology-ochratoxin A. *Veterinary Science Communications,* 1, 349-358.

Gautier, J. C.; Holzhaeuser, D.; Markovic, J.; Gremaud, E.; Schilter, B. & Turesky, R. J. (2001b). Oxidative damage and stress response from ochratoxin A exposure in rats. *Free Radical Biology and Medicine,* 30, 1089-1076.

Gautier, J. C.; Richoz, J.; Welti, D. H.; Markovic, J.; Gremaud, E.; Guengerich, F. P. & Turesky, R. J. (2001a). Metabolism of ochratoxin A: Absence of formation of genotoxic derivatives by human and rat enzymes. *Chemical Research in Toxicology,* 14, 34-45.

Gekle, M. & Silbernagl, S. (1996). Renal toxicodynamics of ochratoxin A: a pathophysiological approach. *Kidney and Blood Pressure Research,* 19, 225-235.

Gekle, M.; Sauvant, C. & Schwerdt, G. (2005). Ochratoxin A at nanomolar concentrations: a signal modulator in renal cells. *Molecular Nutrition & Food Research,* 49(2), 118-130.

Gekle, M.; Schwerdt, G.; Freudinger, R.; Mildenberger, S.; Wilflingseder, D.; Pollack, V.; Dander, M. & Schramek, H. (2000). Ochratoxin A induces JNK activation and apoptosis in MDCK-C7 cells at nanomolar concentrations. *Journal of Pharmacology and Experimental Therapeutics,* 293, 837-844.

Gingras, A. C.; Gygi, S. P.; Raught, B.; Polakiewicz, R. D.; Abraham, R. T.; Hoekstra, M. F.; Aebersold, R. & Sonenberg, N. (1999). Regulation of 4E-BP1 phosphorylation: a novel two-step mechanism. *Genes & Development,* 13(11), 1422-1437.

Green, R. M. & Flamm, S. (2002). AGA technical review on the evaluation of liver chemistry tests. *Gastroenterology,* 123, 1367-1384.

Greenamyre, J. T.; MacKenzie, G.; Peng, T. I. & Stephans, S. E. (1999). Mitochondrial dysfunction in Parkinson's disease. *Biochemical Society Symposia,* 66, 85-97.

Gupta, R. C. (2012). Ochratoxins and citrinin. In: *Veterinary Toxicology: Basic and Clinical Principles* (p. 999). Academic Press.

Hagelberg, S.; Hult, K. & Fuchs, R. (1989). Toxicokinetics of ochratoxin A in several species and its plasma-binding properties. *Journal of Applied Toxicology,* 9, 91-96.

Hasinoff, B. B.; Rahimtula, A. D. & Omar, R. F. (1990). NADPH-cytochrome-P-450 reductase promoted radical production by the iron (III)-ochratoxin complex. *Biocemica et Biophysica Acta,* 1036, 78-81.

Hatey, F. & Galtier, P. (1977). Short term toxicity of ochratoxin A production by *Aspergillus carbonarious* and *A. niger* isolates and detection using coconut cream agar. *Journal of Food Mycology,* 1, 67-72.

Hibi, D.; Kijima, A.; Kuroda, K.; Suzuki, Y.; Ishii, Y.; Jin, M.; Nakajima, M.; Sugita-Konishi, Y.; Yanai, T.; Nohmi, T.; Nishikawa, A. & Umemura, T.

(2013). Molecular mechanisms underlying ochratoxin A-induced genotoxicity: global gene expression analysis suggests induction of DNA double-strand breaks and cell cycle progression. *The Journal of Toxicological Sciences,* 38(1), 57-69.

Hibi, D.; Suzuki, Y.; Ishii, Y.; Jin, M.; Watanabe, M.; Sugita-Konishi, Y.; Yanai, T.; Nohmi, T.; Nishikawa, A. & Umemura, T. (2011). Site-specific *in vivo* mutagenicity in the kidney of *gpt* delta rats given a carcinogenic dose of ochratoxin A. *Toxicological Sciences,* 122, 406-414.

Hood, R. D.; Naughton, M. J. & Hayes, W. A. (1976). Prenatal effects of ochratoxin A in hamsters. *Teratology,* 13, 11-14.

I. A. R. C. (1993). Some naturally occurring substances: Food items and constituents, heterocyclic aromatic amines and mycotoxins. Lyon: World Health Organization.

Iha, M. H. & Trucksess, M. W. (2010). Aflatoxins and ochratoxin A in tea prepared from naturally contaminated powdered ginger. *Food Additives and Contaminants Part A-Chemistry, Analysis, Control, Exposure & Risk Assessment,* 27, 1142-1147.

J. E. C. F. A. (2001). Ochratoxin A. In: *Safety Evaluation of Certain Mycotoxins in Food* (p. 366). Geneva, Switzerland: World Health Organization (WHO). WHO Food Additives Series 47.

Jorgensen, K. (2005). Occurence of ochratoxin A in commodities and processed food: a review of EU occurence data. *Food Additives & Contaminants,* 22, 26-30.

Kaelin Jr, W. G. (2003). The von Hippel-Lindau gene, kidney cancer, and oxygen sensing. *Journal of the American Society of Nephrology,* 14(11), 2703-2711.

Kensler, T. W.; Wakabayashi, N. & Biswal, S. (2007). Cell survival responses to environmental stresses via the Keap1-Nrf2-ARE pathway. *Annual Review of Pharmacology and Toxicology,* 47, 89-116.

Khan, S.; Martin, M.; Bartsch, H. & Rahimtula, A. D. (1989). Perturbation of liver microsomal calcium homeostasis by ochratoxin A. *Biochemical Pharmacology,* 38, 67.

Kim, W. & Kaelin Jr, W. G. (2003). The von Hippel-Lindau tumor suppressor protein: new insights into oxygen sensing and cancer. *Current Opinion in Genetics & Development,* 13(1), 55-60.

Klaunig, J. E. & Kamendulis, L. M. (2004). The role of oxidative stress in carcinogenesis. *Annual Review of Pharmacology and Toxicology,* 44, 239-267.

Krogh, P. (1980). Ochratoxins: Occurence, biological effects and casual role in disease. In: D., Eaker & T., Wadstrom (Eds.), *Natural Toxins* (pp. 673-680). Pergamon Press, Oxford.

Kuiper-Goodman, T. & Scott, P. M. (1989). Risk assessment of the mycotoxin ochratoxin A. *Biomedical and Environmental Sciences,* 2, 179-248.

Kumagai, S. & Aibara, K. (1982). Intestinal absorption and secretion of ochratoxin A in the rat. *Toxicology and Applied Pharmacology,* 64, 94-102.

Lebrun, S. & Follmann, W. (2002). Detection of ochratoxin A induced DNA damage in MDCK cells by alkaline single cell gel electrophoresis (comet essay). *Archives of Toxicology,* 75, 734-741.

Li, J.; Yin, S.; Dong, Y.; Fan, L. & Hu, H. (2011). p53 activation inhibits ochratoxin A-induced apoptosis in monkey and human kidney epithelial cells via suppression of JNK activation. *Biochemical and Biophysical Research Communications,* 411, 458-463.

Liang, R.; Shen, X. L.; Zhang, B.; Li, Y.; Xu, W.; Zhao, C.; Luo, Y.; & Huang, K. (2015). Apoptosis signal-regulating kinase 1 promotes Ochratoxin A-induced renal cytotoxicity. *Scientific Reports,* 5, 8078.

Limonciel, A. & Jennings, P. (2014). A review of the evidence that ochratoxin A is an Nrf2 inhibitor: Implications for nephrotoxicity and renal carcinogenicity. *Toxins,* 6, 371-379.

Liu, J.; Wang, Y.; Cui, J.; Xing, L.; Shen, H.; Wu, S.; Lian, H.; Wang, J.; Yan, X. & Zhang, X. (2012). Ochratoxin A induces oxidative DNA damage and G1 phase arrest in human peripheral blood mononuclear cells *in vitro. Toxicology Letters,* 211, 164-171.

Malaveille, C.; Brun, G. & Bartsch, H. (1994). Structure-activity studies in *E. coli* strains on ochratoxin A (OTA) and its analogues implicate a genotoxic free radical and a cytotoxic thiol derivative as reactive metabolites. *Mutation Research,* 307, 141-147.

Mally, A.; Pepe, G.; Ravoori, S.; Fiore, M.; Gupta, R. C.; Dekant, W. & Mosesso, P. (2005). Ochratoxin A causes DNA damage and cytogenetic effects but no DNA adducts in rats. *Chemical Research in Toxicology,* 18, 1253-1261.

Mally, A.; Zepnik, H.; Wanek, P.; Eder, E.; Dingley, K.; Ihmels, H.; Volkel, W. & Dekant, W. (2004). Ochratoxin A: lack of formation of covalent DNA adducts. *Chemical Research in Toxicology,* 17(2), 234-242.

Marin-Kuan, M.; Nestler, S.; Verguet, C.; Bezencon, C.; Piguet, D.; Mansourian, R.; Holzwarth, J.; Grigorov, M.; Delatour, T.; Mantle, P.; Cavin, C. & Schilter, B. (2006). A toxicogenomics approach to identify new plausible epigenetic mechanisms of ochratoxin A carcinogenicity in rat. *Toxicological Sciences,* 89(1), 120-134.

Marquardt, R. R. & Frohlich, A. A. (1992). A review of recent advances in understanding ochratoxicosis. *Journal of Animal Science,* 70, 3968-3988.

Mayura, K.; Hayes, A. W. & Berndt, W. O. (1983). Effects of dietary protein on teratogenicity of ochratoxin A in rats. *Toxicology,* 27, 147-157.

Monnet-Tschudi, F.; Sorg, O.; Honegger, P.; Zurich, M. G.; Huggett, A. C. & Schilter, B. (1997). Effects of the naturally occurring food mycotoxin ochratoxin A on brain cells in culture. *Neurotoxicology,* 18(3), 831-839.

Muller, G.; Rosner, H.; Rohrmann, B.; Erler, W.; Geschwend, G.; Grafe, U.; Burkert, B.; Moller, U.; Diller, R.; Sachse, K. & Kohler, H. (2003). Effects of the mycotoxin ochratoxin A and some of its metabolites on the human cell line THP-1. *Toxicology,* 184(1), 69-82.

O'Brien, E. & Dietrich, D. R. (2005). Ochratoxin A: the continuing enigma. *Critical Reviews in Toxicology,* 35(1), 33-60.

Obrecht-Pflumio, S.; Chassat, T.; Dirheimer, G. & Marzin, D. (1999). Genotoxicity of ochratoxin A by *Salmonella* mutagenicity test after bioactivation by mouse kidney microsomes. *Mutation Research,* 446(1), 95.

O'Brien, E.; Prietz, A.; & Dietrich, D. R. (2005). Investigation of the teratogenic potential of ochratoxin A and B using the FETAX system. *Birth Defects Research (Part B),* 74: 417-423.

Oka, H.; Chatani, Y.; Hoshino, R.; Ogawa, O.; Kakehi, Y.; Terachi, T.; Okada, Y.; Kawaichi, M.; Kohno, M. & Yoshida, O. (1995). Constitutive activation of mitogen-activated protein (MAP) kinases in human renal cell carcinoma. *Cancer Research,* 55(18), 4182-4187.

Omar, R. F.; Gelboin, H. V. & Rahimtula, A. D. (1996). Effect of cytochrome P450 induction on the metabolism and toxicity of ochratoxin A. *Biochemical Pharmacology,* 51**,** 207-216.

Orlicky, D. J. (1996). Negative regulatory activity of a prostaglandin F2 alpha receptor associated protein (FPRP). *Prostaglandins, Leukotrienes and Essential Fatty Acids,* 54(4), 247-259.

Ozcan, Z.; Gul, G. & Yaman, I. (2014). Ochratoxin A activates opposing c-MET/PI3K/Akt and MAPK/ERK 1-2 pathways in human proximal tubule HK-2 cells. *Archives of Toxicology*, DOI: 10.1007/s00204-014-1311-x.

Patial, V. (2011). Studies on experimental ochratoxicosis and its protection by seabuckthorn (*Hippophae sp.*) in Japanese quail. *Thesis,* p. 64.

Patial, V.; Asrani, R. K.; Patil, R. D.; Ledoux, D. R. & Rottinghaus, G. E. (2013). Pathology of ochratoxin A induced nephrotoxicity in Japanese quail and its protection by seabuckthorn (*Hippophae rhamnoides* L.). *Avian Diseases,* 57, 767-779.

Patil, R. D.; Dwivedi, P. & Sharma, A. K. (2006). Critical period and minimum single oral dose of ochratoxin A for inducing developmental toxicity in pregnant Wistar rats. *Reproductive Toxicology,* 22, 679-687.

Petrik, J.; Zanic-Grubisic, T.; Barisic, K.; Pepeljnjak, S.; Radic, B.; Ferencic, Z. & Cepelak, I. (2003). Apoptosis and oxidative stress induced by ochratoxin A in rat kidney. *Archives of Toxicology,* 77, 685-693.

Petroulakis, E. & Wang, E. (2002). Nerve growth factor specifically stimulates translation of eukaryotic elongation factor 1A-1 (eEF1A-1) mRNA by recruitment to polyribosomes in PC12 cells. *The Journal of Biological Chemistry,* 277(21), 18718-18727.

Pfohl-Leskowicz, A. & Castegnaro, M. (2005). Further arguments in favour of direct covalent binding of ochratoxin A after metabolic biotransformation. *Food Additives & Contaminants,* 22(1), 75-87.

Pfohl-Leszkowicz, A. & Manderville, R. A. (2007). Ochratoxin A: An overview on toxicity and carcinogenicity in animals and humans. *Molecular Nutrition & Food Research,* 51, 61-99.

Pfohl-Leszkowicz, A.; Chakor, K.; Creppy, E. E. & Dirheimer, G. (1991). DNA-adduct formation in mice treated with ochratoxin A. In: M., Castegnaro; R., Plestina; G., Dirheimer; I. N., Chernozemsky & H., Bartsch (Eds.), *Mycotoxins, Endemic Nephropathy and Urinary Tract Tumours* (pp. 245-253). IARC Scientific Publications No. 115, Lyon, France.

Pfohl-Leszkowicz, A.; Grosse, Y.; Kane, A.; Creppy, E. E. & Dirheimer, G. (1993). Differential DNA adduct formation and disappearance in three mouse tissues after treatment with the mycotoxin ochratoxin A. *Mutation Research,* 289, 265-273.

Pinton, P.; Giorgi, C.; Siviero, R.; Zecchini, E. & Rizzuto, R. (2008). Calcium and apoptosis: ER-mitochondria Ca^{2+} transfer in the control of apoptosis. *Oncogene,* 27, 6407-6418.

Polenakovic, M. H. & Steffen, R. (2001). In: A. M., Davison (Ed.), *Oxford Textbook of Clinical Nephropathology* (2, pp. 1203-1211).

Prescott, S. M. & Fitzpatrick, F. A. (2000). Cyclooxygenase 2 and carcinogenesis. *Biochimica et Biophysica Acta,* 1470, 69-78.

Qi, X. Z.; Yang, X.; Chen, S. Y.; He, X. Y.; Dweep, H.; Guo, M. Z.; Cheng, W. H.; Xu, W. T.; Luo, Y. B.; Gretz, N.; Dai, Q. & Huang, K. L. (2014).

Ochratoxin A induced early hepatotoxicity: new mechanistic insights from microRNA, mRNA and proteomic profiling studies. *Scientific Reports,* 4, 5163.

Rached, E.; Hard, G. C.; Blumbach, K.; Weber, K.; Draheim, R.; Lutz, W. K.; Ozden, S.; Steger, U.; Dekant, W. & Mally, A. (2007). Ochratoxin A: 13-week oral toxicity and cell proliferation in male F344/N rats. *Toxicological Sciences,* 97, 288-298.

Ribelin, W. E.; Fukushima, K. & Still, P. (1978). The toxicity of ochratoxin A to ruminants. *Canadian Journal of Comparative Medicine,* 42, 172-176.

Ringot, D.; Chango, A.; Schneider, Y. J. & Larondelle, Y. (2006). Toxicokinetics and toxicodynamics of ochratoxin A, an update. *Chemico-Biological Interactions,* 159, 18-46.

Robbins, S. L. & Angell, M. (1976). *Basic Pathology,* 8th Ed., Saunders Company, Philadelphia, London and Toronto.

Rossiello, M. R.; Rotunno, C.; Coluccia, A.; Carratu, M. R.; Santo, A. D.; Evangelista, V.; Semeraro, N. & Colucci, M. (2008). Ochratoxin A inhibits the production of tissue factor and plasminogen activator inhibitor-2 by human blood mononuclear cells: Another potential mechanism of immune-suppression. *Toxicology and Applied Pharmacology,* 229, 227-231.

Rumora, L. & Grubisic, T. Z. (2009). A journey through mitogen-activated protein kinase and ochratoxin A interactions. *Arhiv za Higijenu Rada i Toksikologiju,* 60: 449-456.

Sauvant, C.; Holzinger, H. & Gekle, M. (2005a). Proximal tubular toxicity of ochratoxin A is amplified by simultaneous inhibition of the extracellular signal-regulated kinases 1/2. *The Journal of Pharmacology and Experimental Therapeutics,* 313(1), 234-241.

Sauvant, C.; Holzinger, H. & Gekle, M. (2005b). The nephrotoxin ochratoxin A induces key parameters of chronic interstitial nephropathy in renal proximal tubular cells. *Cellular Physiology and Biochemistry,* 15(1-4), 125-134.

Sava, V.; Reunova, O.; Velasquez, A.; Harbison, R. & Sanchez-Ramos, J. (2006). Acute neurotoxic effects of the fungal metabolite ochratoxin A. *Neurotoxicology,* 27, 82-92.

Schaaf, G. J.; Nijmeijer, S. M.; Maas, R. F.; Roestenberg, P.; de Groene, E. M. & Fink-Gremmels, J. (2002). The role of oxidative stress in the ochratoxin A-mediated toxicity in proximal tubular cells. *Biochimica et Biophysica Acta,* 1588(2), 149-158.

Schilter, B.; Marin-Kuan, M.; Delatour, T.; Nestler, S.; Mantle, P. & Cavin, C. (2005). Ochratoxin A: Potential epigenetic mechanisms of toxicity and carcinogenicity. *Food Additives & Contaminants,* 22, 88-93.

Schramek, H.; Wilflingseder, D.; Pollack, V.; Freudinger, R.; Mildenberger, S. & Gekle, M. (1997). Ochratoxin A-induced stimulation of extracellular signal-regulated kinases 1/2 is associated with Madin-Darby canine kidney-C7 cell differentiation. *Journal of Pharmcology and Experimental Therapeutics,* 283, 1460-1468.

Schwerdt, G.; Freudinger, R.; Silbernagl, S. & Gekle, M. (1999). The nephrotoxin ochratoxin A induces apoptosis in cultured human proximal tubule cells. *Cell Biology and Toxicology,* 15, 405-415.

Solcan, C.; Coman, I.; Solcan, G.; Miron, L. & Oprean, O. Z. (2008). Histological and ultrastructural lesions of the kidneys in experimental ochratoxicosis of broiler chickens. *Bulletin of University of Agricultural Sciences and Veterinary Medicine Cluj-Napoca Veterinary Medicine,* 65, 91-95.

Sostaric, B. & Vukelic, M. (1991). Characteristics of urinary tract tumours in the area of Balkan endemic nephropathy in Croatia. *IARC Scientific Publications,* 115, 29-35.

Sridhar, S. S.; Hedley, D. & Siu, L. L. (2005). Raf kinase as a target for anticancer therapeutics. *Molecular Cancer Therapeutics,* 4(4), 677-685.

Stemmer, K.; Ellinger-Ziegelbauer, H.; Ahr, H. J. & Dietrich, D. R. (2009). Molecular characterization of preneoplastic lesions provides insight on the development of renal tumors. *American Journal of Pathology,* 175(4), 1686-1698.

Steyn, P. S.; Vleggaar, R.; Du Preez, N. P.; Blyth, A. A. & Seegers, J. C. (1975). The *in vitro* toxicity of analogs of ochratoxin A in monkey kidney epithelial cells. *Toxicology and Applied Pharmacology,* 32, 198-203.

Stojkovic, R.; Hult, K.; Gamulin, S. & Plestina, R. (1984). High affinity binding of ochratoxin A to plasma constituents. *Biochemistry International,* 9, 33-38.

Sujani, N. G. & Sriraman, P. K. (2000). Pathology of ochratoxicosis in quails. *The Indian Journal of Animal Sciences,* 70, 1120-1122.

Takahashi, M.; Papavero, V.; Yuhas, J.; Kort, E.; Kanayama, H. O.; Kagawa, S.; Baxter, R. C.; Yang, X. J.; Gray, S. G. & Teh, B. T. (2005). Altered expression of members of the IGF-axis in clear cell renal cell carcinoma. *International Journal of Oncology,* 26(4), 923-931.

Turski, L. & Turski, W. A. (1993). Towards an understanding of the role of glutamate in neurodegenerative disorders: energy metabolism and neuropathology. *Experientia,* 49, 1064-1072.

Ueno, Y. (1991). Biochemical mode of action of mycotoxins. In: J. E., Smith & R. S., Henderson (Eds.), *Mycotoxins and Animal Foods* (pp. 437-453). CRC Press, Boca Raton, FL.

Vanlangenakker, N.; Berghe, T. V.; Krysko, D. V.; Festjens, N. & Vandenabeele, P. (2008). Molecular mechanisms and pathophysiology of necrotic cell death. *Current Molecular Medicine,* 8(3), 207-220.

Vettorazzi, A.; Gonzalez-Penas, E. & Lopez de Cerain, A. (2014). Ochratoxin A kinetics: A review of analytical methods and studies in rat model. *Food and Chemical Toxicology,* 72, 273-288.

Vogel, E. W. & Nivard, M. J. M. (1993). Performance of 181 chemicals in a Drosophila assay predominantly monitoring interchromosomal mitotic recombination. *Mutagenesis,* 8, 57-81.

W. H. O. (2001). Safety evaluation of certain mycotoxins in food. Fifty-sixth report of the Joint FAO/WHO Expert Committee on Food Additives. In: *WHO Food Additives Series,* No.47.

Wangikar, P. B.; Dwivedi, P. & Sinha, N. (2004). Teratogenic effects of ochratoxin A in rabbits. *World Rabbit Science,* 12, 159-171.

Wei, X. & Sulik, K. K. (1993). Pathogenesis of craniofacial and body wall malformations induced by ochratoxin A in mice. *American Journal of Medical Genetics,* 47(6), 862-871.

Wei, Y. H.; Lu, C. Y.; Lin, T. N. & Wei, R. D. (1985). Effect of ochratoxin A on rat liver mitochondrial respiration and oxidative phosphorylation. *Toxicology,* 36, 119-130.

Xiao, H.; Madhyastha, S.; Marquardt, R. R.; Li, S., Vodela, J. K.; Frohlich, A. A. & Kemppainen, B. W. (1996). Toxicity of ochratoxin A, its opened lactone form and several of its analogs: structure-activity relationships. *Toxicology and Applied Pharmacology,* 137(2), 182-192.

Zamzami, N.; Marchetti, P.; Castedo, M.; Decaudin, D.; Macho, A.; Hirsch, T.; Susin, S. A.; Petit, P. X.; Mignotte, B. & Kroemer, G. (1995). Sequential reduction of mitochondrial transmembrane potential and generation of reactive oxygen species in early programmed cell death. *The Journal of Experimental Medicine*, 182, 367-377.

Zepnik, H.; Volkel, W. & Dekant, W. (2003). Toxicokinetics of the mycotoxin ochratoxin A in F 344 rats after oral administration. *Toxicology and Applied Pharmacology,* 192(1), 36-44.

Zha, S.; Yegnasubramanian, V.; Nelson, W.; Isaacs, W. & De Marzo, A. (2004). Cyclooxygenases in cancer: Progress and perspectives. *Cancer Letters,* 251, 1-20.

Zlender, V.; Breljak, D.; Ljubojevic, M.; Flajs, D.; Balen, D.; Brzica, H.; Domijan, A. M.; Peraica, M.; Fuchs, R.; Anzai, N. & Sabolic, I. (2009). Low doses of ochratoxin A upregulate the protein expression of organic anion transporters Oat1, Oat2, Oat3and Oat5 in rat kidney cortex. *Toxicology and Applied Pharmacology,* 239, 284-296.

In: Ochratoxins
Editor: Daniel Porter
ISBN: 978-1-63483-693-7

Chapter 4

Prevention of Ochratoxin A Production by Using Food Grade Antioxidants

Carla Lorena Barberis, Carina Elizabeth Magnoli * **
and Ana María Dalcero

Departamento de Microbiología e Inmunología,
Facultad de Ciencias Exactas,
Físico- Químicas y Naturales,
Universidad Nacional de Río Cuarto, Córdoba, Argentina

Abstract

Ochratoxin A (OTA) is the second mycotoxin, followed by aflatoxins, of significance in several agricultural products. During the last decades, OTA has been receiving considerable attention worldwide because of its nephrotoxic, immunotoxic, teratogenic and carcinogenic effects on animals. This toxin was included in the group 2B by the International Agency for Research on Cancer (IARC) since it is a potential carcinogen for humans. *Penicillium verrucosum* is the main

* Address correspondence to: Dr. Carina Magnoli, Dr. Carla Barberis. Departamento de Microbiología e Inmunología. Facultad de Ciencias Exactas, Físico, Químicas y Naturales, Universidad Nacional de Río Cuarto, Ruta Nacional N° 36 Km 601, (5800) Río Cuarto, Córdoba, Argentina. Phone and Fax: 54-358-4676231. E-mail: cmagnoli@exa.unrc.edu.ar; E-mail: cbarberis@exa.unrc.edu.ar.

OTA-producing species in temperate and cold climates, and *A. ochraceus* and related species in warm and tropical climates. Due to their physiological differences, each of these species occupies a specific ecological niche. Recently, several studies reported on the presence of potential OTA-producing *Aspergillus* section *Nigri* species in food and feeds. *Aspergillus carbonarius* is considered to be the main producer of OTA, followed by species belonging to the *A. niger* aggregate. The presence of this toxin has been extensively reported worldwide in several products such as coffee, wines, beers, grapes, dried grapes, cereals and derivatives destined for humans and animals, oilseeds and derived products and, occasionally, in body fluids, plasm, meat and kidneys of several animal species. One strategy to reduce the entry of mycotoxins into the food chain is the use of chemical treatments during the storage of cereals, grains and oilseeds to reduce fungal growth and production of toxins. From a human health perspective, the use of antioxidants such as butylated-hydroxyanisole (BHA), propyl paraben (PP) and butylated-hydroxytoluene (BHT) are allowed by the US Food and Drug Administration (FDA) as antimicrobial agents. Such compounds are regarded as safe (GRAS) chemicals. The last known studies have shown that these antioxidants have a protective action in food since they could maintain its organoleptic properties. This chapter reviews data showing that treatments with food grade antioxidants significantly influence growth parameters, enzymes production and OTA production of ocratoxigenic strains. In addition, the incubation time, temperature, water activity and their interactions are also important factors involved in the development of these fungal species and in the accumulation of toxins.

INTRODUCTION

Ochratoxin A (OTA) $C_2OH_{18}CINO_6$ is a polyketide mycotoxin composed of a chlorinated dihydro-methyl-isocoumarin moiety linked via an amide bond to phenylalanine [1]. OTA has been shown to be a potent nephrotoxic mycotoxin linked to kidney dysfunctions in both humans and animals. OTA has also been associated to carcinogenic, genotoxic and immunotoxic effects. During the last decade, OTA has been receiving considerable attention worldwide because of the hazard that it represents to the health of humans and animals [2]. Due to its toxicity, the International Agency for Research on Cancer [3] has classified OTA into the group 2B (potential carcinogen to humans).

The natural occurrence of this mycotoxin has been informed worldwide in several commodities as cereals, oilseeds, grapes, coffee, cocoa, speces, poultry, pork and their derived products [4]. Due to the extensive presence of this toxin in a wide variety of foodstuffs, maximum permitted levels have been established by the EU and other countries to reduce the risk of human exposure [5].

Although production of OTA was attributed to *Aspergillus ochraceus* in the first studies, several species from the genera *Aspergillus* and *Penicillium* are currently being reported as producers of this mycotoxin. Contamination of food and feed with OTA was related, until recently, to *A. ochraceus* and *P. verrucosum*. These fungi affect mainly dried stored foods and cereals, respectively, throughout different regions of the world. Recently, several studies have shown that *A. carbonarius* and other species of *Aspergillus* section *Nigri* are the main producers of OTA in several agricultural commodities from warm and tropical areas [6-10].

The temperature and the moisture content are the main factors related to fungal growth and the consequent production of OTA during cultivation, harvesting and storage. Other important factors involved are the presence of fungal spores, the mechanical damage of the crop, the presence of insects, the pH, the oxygen and carbon dioxide levels, chemical and physical treatments, and their interaction [11]. Many strategies have been developed to decrease contamination by OTA. Prevention through pre-harvest management practices is the best method for controlling contamination. However, if contamination occurs after this stage, the OTA derived hazards must be managed through post-harvest strategies. Storage is one of the most critical post-harvest stages during handling of food. Maintaining agricultural commodities below the optimal water activity and temperature for fungal growth limit the damage by these organisms [11, 12]. However, this aspect is not sufficient for preventing spoilage and production of the mycotoxin during storage. Currently, the use of chemical products and natural agents constitute alternatives for the control of fungal growth and production of OTA during storage. During the last years, food grade antioxidants have been used as an alternative to traditional fungicides to safely control fungal growth in several food and agricultural products.

Several phenolic antioxidants showed biocidal action *in vitro* against yeasts [13] and filamentous fungi [14]. These compounds showed capacity to control the growth of mycotoxigenic fungi and the accumulation of the mycotoxin in synthetic media and agricultural products such as corn and peanut kernels [15- 26]. In this chapter, we review the main data on the effects of phenolic antioxidants on fungal growth and production of OTA.

Toxicological Assessment of Ochratoxin A

Exposure of humans to toxic compounds is associated with several health disorders such as reproductive alterations and birth defects, immunological suppression, kidney and liver dysfunction, mental health problems and cancer [2]. Mycotoxins are considered as highly dangerous substances to humans and they represent the larger group of toxins that caused alert notifications in the European Rapid Alert System for Food and Feed (RASFF) during 2011 [27]. Recently, the chemical structures of diverse ochratoxins and their metabolites were elucidated [28]. Undoubtedly, ochratoxin A (OTA) is the mycotoxin that has received the most attention due to the risk that it represents to human and animal health. Several studies have informed on its nephrotoxic, hepatotoxic, teratogenic, immunotoxic, neurotoxic, genotoxic and carcinogenic effects [29-32].

Although OTA accumulates in several organs, the kidneys represent the main targets. Inside these organs, it exerts toxic and carcinogenic effects [29]. Three renal diseases have been associated to exposition to OTA. The most common is known as Balkan endemic nephropathy (BEN), followed by the chronic interstitial nephritis and the karyomegalic interstitial nephritis. BEN is a chronic nephropathy reported in several rural regions from Bulgaria, Romania, Serbia, Croatia, and Bosnia [33]. Tubulointerstitial chronic nephritis, urotheliomas and renal carcinoma have also been reported in patients with BEN [34]. Although considerable research has been done, the physiological role of OTA in patients with BEN has not yet been completely determined [35].

The International Agency for Research on Cancer [3] evaluated the experimental evidence for carcinogenicity as sufficient and classified OTA as "possibly carcinogenic to humans" (group 2B), but not yet "probably carcinogenic to humans" (group 2A). However, more recent studies have reported additional data in favor of a classification into group 2A [35].

Producers of Ochratoxin A

This toxin was originally described as a metabolite from *Aspergillus ochraceus* growing in a pure culture [36]. The latest information indicates that *P. verrucosum* is the main species associated to production of OTA in foods and feeds in temperate climates. Several studies reported that OTA is produced as a secondary metabolite by several toxigenic fungal species belonging to the genera *Aspergillus* section *Nigri* and *Circumdati* (*A. carbonarius*, *A. foetidus*, *A. lacticoffeatus*, *A. ochraceus*, *A. niger*, *A. sclerotioniger*, *A. steynii*, *A. tubingensis*, *A. westerdijkiae*) and *Penicillium* (*P. nordicum* and *P. verrucosum,* among others). In addition, members of *Aspergillus* section *Flavi*, including *Petromyces albertensis* and *P. alliaceu*, have been reported as producers of OTA [6-10, 37-39].

OTA-producing species of *Penicillium* are related to temperate and cold climates, while most of the information on *Aspergillus* species comes from warm and tropical climates. *P. verrucosum* is mostly found contaminating plants in fields prior to harvest. On the other hand, *P. nordicum* is associated with cheese, meat and meat by products [40]. During the last years, several studies reported on the presence of potential OTA-producing *Aspergillus* section *Nigri* species in foods and feeds such as cocoa, coffee beans, peanuts, dried grapes, grapes, wines and poultry feed, among others. In these studies, *Aspergillus carbonarius* is considered to be the main producer of OTA followed by species belonging to *A. niger* aggregate. A relation between this mycotoxin, the substrate and fungal ochratoxigenic species has been informed. For instance, OTA is mainly produced by *P. verrucosum* in cereals. By comparison, OTA is produced by *A. carbonarius* and related species of this section in coffee, cocoa, wine, grapes and dried vine fruits from main viticultural regions worldwide [6, 41-45].

Similarly, extensive information is available worldwide on the presence of OTA in a wide range of foodstuffs of both plant and animal origin, such as coffee, cocoa [12, 46-48], wines [49, 50], beers [51-53], grapes [54-57], dried grapes [48, 58-60], cereals and derivatives destined to humans and animals [61-65], oilseeds and their byproducts [61, 62, 66] and, occasionally, in body fluids, plasm, meat and kidneys of several animal species [67-70]. The maximum OTA levels in these products are strongly regulated by the European Union [5].

SIGNIFICANCE OF OCHRATOXINS IN FOOD

Each year an important percentage of crops are contaminated with mycotoxins, among them OTA, which generates health risks to consumers and incomes losses to farmers. Some strategies are applied to prevent the entry of these metabolites in the food chain, to protect the public health and to reduce the economic losses caused by contaminated agricultural products.

Ochratoxins are chemically stable and do not seem to be significantly affected by processing temperatures applied to the products [71]. For instance, Puntaric et al. [72] found that the baking and roasting processes only reduce the toxin content in approximately 20%. On the other hand, boiling has no effect on OTA. Thus, the knowledge of the ecophysiology of OTA producing fungi and of the ecological factors that influence on production of OTA is essential to optimize the implementation of preventive strategies. The goal is to control the sanitary quality of raw materials and/or products that are susceptible to fungal colonization.

FACTORS INVOLVED IN PRODUCTION OF OCHRATOXIN A

There are multiple factors significantly involved in the development of ochratoxigenic species and in the biosynthesis of secondary metabolites in food and feeds. Among these, water activity (a_W) and oxygen availability, temperature and pH have shown the highest effects on growth and production of OTA. In general, the toxigenic species are not aggressive pathogens. However, they are often well adapted to substrates with low humidity, and they can easily colonize stored crops when environmental conditions are appropriate. In the last decades, the ecophysiological studies have focused on the main OTA producing species *P. verrucosum* and *A. ochraceus* and on the influence of environmental factors and nutrient sources on growth and production of OTA. The growth parameters vary widely between different *Penicillium* and *Aspergillus* species. Available data indicate that production of OTA is stimulated by a high a_W level [73-81].

Recently, many works have informed on the influence of a_W and temperature on the biosynthesis of OTA by ochratoxigenic species belonging to *Aspergillus* section *Nigri* such as *A. niger* aggregate and *A. carbonarius*. In general, the temperature and the a_W range are more restrictive for production of OTA than for fungal growth, both on natural substrates and culture media [26, 82-88]. Production of OTA in *Aspergillus* generally occurs between 20 and 30ºC, which are temperatures below optimal for fungal growth [73, 76, 89]. With regards to *Penicillium* species, the available data showed that the growth is favored at temperatures below 30ºC and that the synthesis of OTA occurs at lower temperatures (5ºC) [90].

Other important factors involved in production of OTA are the presence of fungal spores, the mechanical damage of crops, the presence of insects, chemical and physical treatments, and their interaction [11, 91]. Additionally, harvesting and storage, which are among the critical factors that affect fungal growth during farming as well as the stages at which products remain under adverse environmental conditions, are determining factors for accumulation of OTA. The knowledge of the ecophysiology of OTA producing strains is critical in the development and prediction of risk models for avoiding contamination of both raw materials and finished foods subjected to the interaction of several environmental parameters.

PREVENTION AND DECONTAMINATION STRATEGIES

Nonetheless, it is important to consider that it is possible to detect OTA in several substrates without the presence of the producing fungal species. This can be explained because chemical processes or environmental changes can inhibit the fungal growth but not alter the synthetized mycotoxin that remains in the substrate [92].

Several critical points of control in food manufacturing need to be considered to prevent production of OTA. For instance, drying, processing and storage of cereals, drying, transport and storage of coffee, cocoa and spices, and harvesting and transport of grapes [12, 89, 93]. A complete removal of OTA from foods is not feasible. Because of this, the Codex Alimentarius Commission [94] established some strategies to achieve reduction of OTA.

One of these strategies is applied during the pre-harvest stage of crops. It consists in the maintenance of the integrity of crops in order to avoid colonization by fungi. In this sense, the mechanical damage and the attack by insects are the most controlled processes. The elimination of contaminated agricultural residues also minimizes the risk of contamination [93]. The harvest of products during their optimal maturation stage is one of the important factors for reducing contamination by OTA. In addition, an optimal moisture content in products helps to avoid the mechanical damages and the collection in sacks or containers free of contamination diminishes the risks of contamination. For example, overripe, fermented, damaged or fallen fruits must be discarded, as they may contain high levels of OTA or harbor OTA-producing fungi that could be rapidly propagated [95]. The use of chemical compounds is other attractive strategy to prevent production of OTA. Results have shown that some fungicides are effective against fungal colonization and production of OTA [55], while others such as carbendazim stimulate production of OTA [96]. In addition, the use of native microorganisms or their products have been proposed as biocontrol strategies [97, 98].

Prevention through pre-harvest management practices is the best method for controlling contamination by mycotoxins. However, if the contamination occurs or persists after this phase, the hazards associated to OTA must be managed through post-harvest procedures. The traditional means to avoid fungal growth in grains is to dry them rapidly after harvest and keep them dry. Pardo et al. [75] suggested that a reduction of the moisture content of grains to provide an a_W below 0.8 is necessary to prevent production of OTA by *Aspergillus* section *Nigri* species.

During post-harvest, storage and processing are the major stages in which contamination can be prevented or in which detoxification can take place by means of a physical separation or chemical inactivation after specific procedures. When stored grains are colonized by fungal species the application of different techniques such as segregation of grains, fumigation, aeration and cooling, sealing of products and control of the atmospheres become necessary. These approaches are applied especially in tropical and subtropical regions where damage during storage is a major problem.

Storage is one of the most critical post-harvest phases during food handling. Thus, it must be carried out under perfectly controlled conditions. For instance, it was observed that grains stored with an a_W of 0.70 or lower (<14.5% moisture by weight) and at temperatures under 20°C [11, 12] will not be subjected to spoilage and accumulation of mycotoxins. Ecophysiological studies with black Aspergilli showed that the use of different storage practices such as a_W and temperature levels below 0.930 and 15°C, respectively, allow to control contamination by fungi and to minimize the production of OTA in several products such as peanuts, corn, dried grapes and derived products for human consumption [83]. Magan and Aldred [99-101] showed that the adding of CO_2 to the atmosphere during the storage of grains allows detoxification of OTA. However, the relative success of this method depends on the ochratoxigenic species. The use of chemical and natural agents constitutes an attractive alternative for the control of OTA during storage. Preservatives such as potassium sorbate or calcium propionate were effective in preventing contamination by OTA in bread [77]. Similarly, Amezqueta et al. [102] showed that sodium bicarbonate and sodium carbonate are capable to decontaminate cocoa shells from OTA under different pressures, temperatures and time conditions.

A diversity of fungicides were tested in their effectiveness to prevent production of OTA. However, it is evident in some species that the use of certain concentrations of these compounds (especially those containing sulfur) produce stress conditions, which could stimulate production of OTA [77, 96, 99-101]. Due to the increasing number of fungal strains that become resistant to chemical fungicides and the impact of pesticides on the environment and on human health, alternative methods have been proposed to substitute or complement treatments with fungicides to control fungi under field or storage conditions.

Among all of the products, the antioxidants vanillic acid and 4-hydroxybenzoic acid and essential oils extracted from plants, among others, have been proposed to inhibit the production of OTA [24, 103, 104]. At the present time, other alternative is the use of yeasts. In this sense, several yeast species from different genera are considered to be potential biocontrol agents to control both growth of ochratoxigenic *Aspergillus* species and accumulation of OTA [105]. Other studied products used to control OTA are natural extracts. Ersilia et al. [106] suggested that the natural extract from grape pomace and grape seeds has fungicidal and fungistatic activities and reduces the accumulation of OTA in wheat, at least similarly to the antioxidant-butylated hydroxytoluene (BHT).

PHENOLIC ANTIOXIDANTS AND GENERAL ASPECTS

One of the main reasons for food deterioration during processing or storage is the oxidation of lipids, which evolves through different steps and originates the formation of very reactive compounds. It must be noted that this degradation process occurs not only in food lipids, but also in those present in cells and biological systems. Antioxidants are substances that markedly delay or prevent the oxidation of the substrate when present in food at low concentrations [107, 108]. Antioxidants are usually classified into two groups: natural and synthetic antioxidants. Synthetic antioxidants are widely used as food additives due to their high effectiveness, low cost and wide availability. Butylated-hydroxytoluene (BHT E-321) and butylated-hydroxyanisole (BHA E-320) are the antioxidants most widely used in foods and in food-related products, such as packaging. The antioxidant properties of BHT and BHA rely on its molecular configuration, since, like other synthetic phenolic antioxidants, it possesses a labile hydrogen atom in the hydroxyl group that can be donated and reduce the free radicals generated during the initiation step of the oxidation of lipids. Thus, this antioxidant itself is oxidized and the subsequent derived radical is stabilized by electronic delocalization in the benzene ring. Consequently, BHT and BHA can stop the propagation of radical oxidation, which delays the oxidation of lipids and extends the shelf life of foods [109]. Phenolic antioxidants have been being added to vegetable oils since they prevent the rancidification of food, a process that creates objectionable odors [110]. However, some studies have proved that the addition of artificial antioxidants may cause a loss of nutritional quality and even produce toxic effects. In fact, the use of these antioxidants is controlled in a number of countries [111] and, consequently, the amounts of these substances in food products are adjusted to prevent undesirable effects.

Parabens are widely used alone or in combination as antimicrobials in personal care products (cosmetics, shampoos, shaving products, lotions) food, food packaging and pharmaceutical products [112]. Propyl paraben (PP) is a stable, non-volatile compound used as an antimicrobial preservative in foods, drugs and cosmetics for over 50 years. PP is an ester of p-hydroxybenzoate that is readily absorbed via the gastrointestinal tract and dermis. Later, it is hydrolyzed to p-hydroxybenzoic acid and conjugated. The conjugates are rapidly excreted in the urine. There is no evidence on its accumulation. Acute toxicity studies in animals indicate that PP is relatively non-toxic by both oral and parenteral routes, although it is mildly irritating to the skin. PP is not carcinogenic, mutagenic or clastogenic. Finally, it is not cytogenic *in vitro* in the absence of carboxyesterase inhibitors [113].

The use of antioxidants and antimicrobial agents such as BHA, PP and BHT is allowed by the US Food and Drug Administration (FDA) since they are considered as safe chemicals (GRAS). In addition, several studies revealed that different phenolic antioxidants such as BHT, BHA and propyl gallate have beneficial effects on human and animal health [114].

In relation to their effect on the environment, it is known that phenolic antioxidants are safe additives for food, whereas traditional fungicides are dangerous for humans and animals. This finding further emphasizes the impact of food grade antioxidants against fungi on stored grains and strengthens the possibility of their use as an alternative to replace the traditional fungicides for controlling storage fungi and occurrence of OTA.

PHENOLIC ANTIOXIDANTS AND CONTROL OF OCHRATOXIN A PRODUCTION

Data on phenolic antioxidant compounds have been reported with coffee [115], grapes and wine [116], barley [117, 118], tree nuts including walnut [119], almond [120, 121], and pistachio [122, 123]. These antioxidant sources are also common habitats for ochratoxigenic *Aspergillus* and *Penicillium* species, which can lead to a potential contamination of these commodities with OTA. This apparent contradiction indicates that the roles of antioxidants and production of OTA are complex in the ecology of ochratoxigenic fungi. Individual antioxidants have variable effects on both production of OTA and growth of individual ochratoxigenic *Aspergillus* strains. These effects may or may not be independent from each other.

Several works have demonstrated that the food grade antioxidant BHA and the antimicrobial PP can control the growth of *Fusarium* and *Aspergillus* section *Flavi* species as well as their production of mycotoxins [15-17, 19-23]. In a previous study, the effect of BHA and PP on the growth rate and production of OTA of *Aspergillus* section *Nigri* species was also evaluated on peanut meal-based medium [124, 125].

STUDIES ON THE EFFECTS OF PHENOLIC ANTIOXIDANTS ON PRODUCTION OF OTA *IN VITRO*

Recently, the effect of BHA and PP alone on the growth rate and production of OTA of *Aspergillus* section *Nigri* species was evaluated on peanut meal extract agar at a wide range of concentrations (1 to 20 mM) and under three water activity conditions [124, 125]. The results suggest that the growth rate and the production of OTA of these strains are completely inhibited by concentrations of 20 and 5 mM of BHA and PP, respectively. The evaluated growth parameters (lag phase and mycelial growth rate) are significantly influenced by a_W, temperature, concentration of BHA and their interactions. With respect to production of OTA, the temperature was the lesser influencing variable among the evaluated factors. High concentrations of BHA (10-20 mM) inhibited the growth and lag phase of *Aspergillus* section *Nigri* species at different a_W levels. Production of OTA followed the same general behavior as those observed with growth parameters. Marin et al. [126, 127] demonstrated that a_W is an important factor that influences growth and production of mycotoxins when different fungal species grow in the presence of antioxidants. Other researchers, who evaluated this antioxidant for the control of *Aspergillus* section *Flavi* species [19, 22, 128-130], observed that BHA caused a significant reduction of the growth rate (between 20 and 74%) and production of aflatoxin B_1 at a concentration of 1 mM. The obtained results with *Aspergillus* section *Nigri* strains showed that these reduction percentages were achieved with lower concentrations of BHA. To our knowledge, only one work evaluated the impact of PP on the growth and production of OTA of *Aspergillus* section *Nigri* species *in vitro* with a_W of 0.995, 0.980 and 0.930 and at 18 and 25ºC [125]. The tested strains were not able to reach the exponential phase with 5, 10 and 20 mM of PP and at all of the evaluated environmental conditions. At 1 mM of the antioxidant, all of the strains showed an increase in their lag phase. The effect of PP on the growth rate of *Aspergillus* section *Nigri* strains at different a_W and temperature levels was also evaluated. Propyl paraben, at 5, 10 and 20 mM, completely suppressed the mycelial growth and the production of OTA of these strains at different environmental conditions. This result was also observed at 1 mM when all of the *Aspergillus* section *Nigri* strains grew at 18ºC. At 25ºC, an important reduction (90 and 76%) in production of OTA was observed in *A. carbonarius* and *A. niger* aggregate strains, respectively. In other study [124] carried out with the same strains, a complete inhibition of growth and production of OTA was observed at highest levels of PP (20 mM) and at all of the environmental conditions tested. These authors concluded that the antioxidant PP is more efficient than BHA in the control of growth and production of OTA of *Aspergillus* section *Nigri* on peanut based media.

In contrast, it has been shown that lower levels of PP (1 mM) cause a stimulation of OTA accumulation over a range of environmental conditions such as a_W of 0.995 and 0.980 and temperature of 18°C [125]. However, at concentrations of PP higher or equal than 5 mM, the production of toxin was completely inhibited. Other authors reported similar results using other antioxidants. Passone et al. [22] evaluated the effect of several antioxidants on the growth and production of aflatoxin by *Aspergillus* section *Flavi* strains. These authors demonstrated that a sub-inhibitory concentration (1 mM) of BHT and propyl-hydroxybenzoate (THB) increased the growth rate. In addition, they observed that BHT, at this concentration, stimulated the production of AFB_1. The observed stimulation suggests that it is essential to determine the optimal antifungal dose to use for each fungal species. High levels of antioxidants could produce undesirable effects on the organoleptic properties of grains and on their processed products, causing a negative economic impact. On the other hand, sub-inhibitory doses together with inadequate distribution of preservatives, especially at low a_W values could cause stimulation of fungal sporulation, growth and secondary metabolism with increasing mycotoxin production [131].

Some authors [25] evaluated the effects of phenolic antioxidants on production of OTA by *A. carbonarius* on a synthetic medium. As a result, they observed that some antioxidants, e.g., gallic, 4-hydroxybenzoic and chlorogenic acids tended to inhibit production of OTA. In other work, Palumbo et al. [24] studied the effect of phenolic antioxidants such as gallic acid, vanillic acid, protocatechuic acid, 4-hydroxybenzoic acid, catechin, caffeic acid, and chlorogenic acid (at 1 to 20 mM) on growth and production of OTA of ochratoxigenic *Aspergillus* species. A minimal medium containing glucose and NH_4Cl as carbon and nitrogen source, respectively, was used. The greatest inhibition of production of OTA was observed in *A. sulphureus*, *A. elegans*, and *A. lanosus*. Vanillic acid and 4-hydroxybenzoic acid strongly inhibited production of OTA and growth of most of the tested strains. However, *A. ochraceus* was not inhibited by either compound. The effects on *A. carbonarius* were highly variable. Although the results were not statistically significant, gallic acid, 4-hydroxybenzoic acid and chlorogenic acid showed a weak inhibition of production of OTA. In contrast, the vanillic, caffeic and protocatechuic acids and catechin produced widely divergent effects among replicate samples, ranging from a 97% inhibition of OTA to 3-fold higher OTA levels compared to the control treatment. It should be noted that very low levels of OTA are produced in the control cultures of this species in comparison to the other tested strains, which can be attributed to the high variability in its response to treatment with antioxidants. Romero et al. [132] reported on the inhibition of the growth and production of OTA of *A. carbonarius* on Czapek Yeast Extract agar supplied with caffeic acid and the flavonoids rutin and quercetin. All of the phenolic compounds caused a significant reduction of growth rate and lag phase at 250 mg L^{-1}. A Higher concentration of gallic acid (500 mg L^{-1}) had a strong effect on the lag phase and growth rate of *A. carbonarius* strains. By comparison, a significant effect on production of OTA was observed even with the lowest concentration (100 mg L^{-1}) used. In general, the effect on growth was less evident than the effect on toxin production.

Many food preservatives and fungicides have been used in combination. Indeed, some works showed that low concentrations of both antioxidants may be more effective together than either one alone. The mixture of antioxidants (BHA and PP) was also proven to be effective in the control of the growth of aflatoxigenic fungi and production of aflatoxins. Based on these data, assays were performed to also explore the antifungal effectiveness of PP in combination with BHA on ochratoxigenic *Aspergillus* section *Nigri* strains [133]. In general, the lag phases of *A. carbonarius* and *A. niger* aggregate strains increased significantly in the presence of the highest concentrations of the combinations of the antioxidant and the antimicrobial compound. The mixture BHA+PP (5.0+2.5 mM) caused the longest lag phases before growth. In addition, all of the studied strains were not able to reach the exponential phase in the presence of the highest concentration of the BHA+PP mixture tested (10+5.0 mM). When growth rate was analyzed, the authors observed that the BHA+PP mixtures of 10+2.5 mM and 5.0+2.5 mM completely suppressed the mycelial growth of all of the *Aspergillus* section *Nigri* strains under the different a_w levels assayed. No significant differences were observed between the mixtures of 0.5+1.0 mM and 1.0+1.0 mM on the growth rate of *A. carbonarius* and between the mixtures of 2.5+0.5 mM and 0.5+1.0 mM for *A. niger* aggregate strains. The effect of the concentration range of BHA+PP/a_w treatments at 25ºC on production of OTA were also tested. In general, production of OTA did not show a pattern similar to that of the growth rate. The effect of the treatments with antioxidants was variable and depended on the species and the a_w assayed. Production of OTA by *A. carbonarius* strains was significantly reduced in the presence of all of the mixtures, compared to controls. On the other hand, the response of *A. niger* aggregate strains to the binary mixture assayed was shown to vary. A significant stimulation of production of OTA was observed with the lowest concentration of the BHA+PP mixture whereas mixtures of 2.5+1.0 mM, 5.0+2.5 mM and 10+2.5 mM completely inhibited the production of OTA in all of the strains. These obtained results show that the combinations of BHA and PP at lower concentrations than those usually applied cause the inability of the *Aspergillus* section *Nigri* strains to reach the exponential phase as well as a complete inhibition of their growth rate and production of OTA. The observed stimulation of OTA with mixtures of BHA+PP at low concentrations suggests that the control of growth does not necessarily implies inhibition of toxin production. In previous works, a stimulation of the production of other mycotoxin was observed in the presence of low concentrations of antioxidants. Stimulation of fumonisin in *F. verticillioides* and *F. proliferatum* was reported with a mixture of BHT+PP (1 mM) with a_w of 0.995 on maize based medium [16]. Similar effects have been reported by Nesci et al. [134], who observed stimulation of AFB_1 production in *A. flavus* and *A. parasiticus* strains with combinations of subinhibitory concentrations of natural phytochemicals (ferulic and cinamic acids) on maize grains under different a_W conditions. It has also been reported that subinhibitory doses together with inadequate distribution of preservatives, especially with low a_W, could enhance fungal growth [135, 136].

Reynoso et al. [16] demonstrated a synergistic effect of the BHA+PP mixture at 0.5 + 1.0mM in the control of the growth rate and production of fumonisin of *F. verticillioides* and *F. proliferatum*. In other study [23], the authors showed that high concentrations of BHA+PP (20+10 or 10+20 mM and 10+20 or 20+10 mM) completely inhibited the growth rate of *A. flavus* and *A. parasiticus* strains on peanut seeds. In addition, all of the antioxidant mixtures used (BHA+PP, 10+10, 10+20, 20+10 and 20+20 mM) significantly reduced accumulation of AFB_1 after 11 and 35 days of experimentation. Moreover, other combinations of antioxidants have been tested as potent fungal inhibitors, and it was observed that benzoic and sorbic acids together are more effective when used in combination than alone, and inhibit spoilage by yeasts and filamentous fungi [137].

Studies *In Situ* on the Effect of Phenolic Antioxidants on Production of OTA

There are few data on the effects of phenolic antioxidants on development and production of OTA of black Aspergilli *in situ*. Barberis et al. [138] evaluated the inhibition of the growth and production of OTA of *Aspergillus* section *Nigri* species by BHA and PP in peanuts at 18 and 25°C and under different a_w conditions. The results showed a significant increase in the length of the lag phase with 10 mM of BHA or PP. The strains did not reach the exponential phase with 20 mM of BHA and at 18°C. Moreover, the authors observed that the strains were not able to reach the exponential phase with 20 mM of PP at 18°C and at any a_W. Butylated hydroxyanisole, at 20 mM and 18°C, completely inhibited the growth of all of the strains. BHA only reduced the growth rate of *A. carbonarius* strains at 18°C and with a concentration of 5 and 10 mM. At 25°C, a significant reduction in the growth rate of all of the strains was observed with 5 and 10 mM of BHA. Propyl paraben completely inhibited the growth of all of the strains at the highest level used (20 mM) and at 18°C. A complete inhibition of *A. carbonarius* growth also occurred with a concentration of 10 mM. At 25°C, the mycelial growth of all of the strains was significantly reduced with 20 mM of PP. On the other hand, BHA at 5, 10 and

20 mM completely inhibited the production of OTA by *A. carbonarius* strains at 18ºC. Also, a significant reduction in the production of OTA by *A. niger* aggregate strains occurred at this temperature with 5 and 10 mM of BHA. In general, BHA did not decrease the production of OTA at 25ºC. In contrast, production of OTA increased at 10 and 20 mM and 0.95 a_w from 47 ng $g^{-1}$to 638 and 1582 ng $g^{-1,}$ respectively. These results showed that different levels of antioxidant or antimicrobial compound stimulate accumulation of OTA in *A. carbonarius* and *A. niger* aggregate strains. This stimulation was also observed with BHA and PP in studies *in vitro*, although with a lower concentration (1 mM) that stimulated both fungal growth and production of OTA in some *Aspergillus* section *Nigri* strains [124-125]. These results were also reported by other authors, using different antioxidants. Passone et al. [22] evaluated the effect of several antioxidants on growth and production of AFB_1 of *Aspergillus* section *Flavi* strains. They found that sub-inhibitory concentrations (1 mM) of BHT and THB increased the fungal growth rate and that BHT stimulated production of AFB_1. In other work, Passone et al. [23] evaluated two concentrations of BHA on production of AFB_1 by *A. flavus* strains growing on peanuts at 28ºC. They observed that 20 mM of this antioxidant completely inhibited toxin production. Similarly, the effects of PP on growth parameters and production of mycotoxin were informed with *Aspergillus* section *Flavi* strains in peanuts. Passone et al. [23] showed that growth inhibition was observed with 10 and 20 mM of this antimicrobial compound and that production of AFB_1 was only significantly reduced with 20 mM.

PHENOLIC ANTIOXIDANTS AND PRODUCTION OF HYDROLITIC ENZIMES

Quantitative enzyme assays for specific hydrolytic enzymes produced by spoilage inducing fungi have been shown to be effective early indicators of growth prior to visible molding [139]. A number of authors [139-141] evaluated the environmental factors affecting production of hydrolytic enzymes by toxigenic fungi during colonization of grains. However, few works have evaluated the influence of these enzymes on OTA producing strains.

Kinetic studies have shown a close correlation between production of hydrolytic enzymes and production of OTA in *P. verrucosum* [100, 101]. The genus *Aspergillus* has been associated with the production of diverse hydrolytic enzymes such as N-acetyl-β-D-glucosaminidase, β-d-glucosidase, α-d-galactosidase, phosphatases, fucosidases, pectinases, lipases, amylases and sulphatases [139-142].

Barberis et al. [143] evaluated the effect of BHA and PP on two of the most important hydrolytic enzymes (β-D-glucosidase and α-D-galactosidase) produced by *Aspergillus* section *Nigri* species under different a_W conditions and incubation times on peanut-based medium. The highest inhibition of the β-D-glucosidase activity of *A. carbonarius* and *A. niger* was found with 20 mmol L^{-1} of BHA or PP and a_W of 0.98 and 0.95, respectively. By comparison, a significant decrease in α-D-galactosidase activity was observed in *A. carbonarius* with 5 to 20 mmol L^{-1} of BHA or PP, under all of the tested conditions. With regards to *A. niger*, the highest percentages of inhibition of enzymatic activity were found after 96 h with 20 mmol L^{-1} of BHA or PP and a_W of 0.95. The effects of BHA or PP on the enzymes involved in the process of early substrate colonization suggest that the ochratoxigenic strains could be kept at the lag phase for a certain period of time by using specific concentrations of BHA and PP, which would delay the spoilage and the contamination of grains by OTA during storage.

ACTIVITY OF PHENOLIC ANTIOXIDANTS

In the literature, several mechanisms have been proposed to explain the efficacy of the antimicrobial proprieties from phenolic antioxidants. Previous evidence has confirmed that they interfere with the structure of the lipid membrane. Specifically, BHA causes cellular leakage in bacteria and increases leakage of carbohydrates, amino acids and proteins from fungal mycelia, which suggests a disruption of the cellular membranes [130, 144]. On the other hand, the mechanism of PP may be linked to a mitochondrial failure dependent on the induction of the transition of membrane permeability accompanied by a mitochondrial depolarization and a depletion of cellular ATP through the uncoupling of oxidative phosphorylation [113].

The mechanism of inhibition of fungal growth and mycotoxin production by PP has not been demonstrated so far. Some authors suggested that the toxicity of phenolic compounds towards fungi is based on the inhibition of fungal enzymes, the destruction of the permeability barrier of the cell membrane and the disruption of the synthesis of nucleic acids [137-145].

OXIDATIVE STRESS AND ANTIOXIDANTS

Oxidative stress is recognized as a trigger of mycotoxin biosynthesis. Jayashree and Subramanyam [146] were among the first researchers who evidenced the relevance of reactive oxygen species (ROS) and oxidative stress on production of mycotoxins. They found that production of aflatoxin is a consequence of increased oxidative stress, leading to enhanced lipid peroxidation and generation of ROS. Similar results were reported based on the study of the accumulation of deoxynivalenol by *Fusarium graminearum*, aflatoxins by *Aspergillus* section *Flavi* and sterigmatocystin by *A. nidulans* [147-149]. By comparison, few works have evaluated the relationship between oxidative stress and biosynthesis of OTA.

Fanelli et al. [150] reported on the effective inhibition of production of OTA by *P. verrucosum* by using the antioxidants resveratrol and BHA. In other study with resveratrol and three essential oils, Aldred et al. [79] informed that the antioxidant resveratrol was more efficient for controlling growth of *P. verrucosum* and *A. westerdijkiae* colonizing wheat grains and production of OTA on irradiated wheat grains, compared to the three essential oils. In general, the growth was relatively unaffected by a low concentration of essential oils (50 μg mL^{-1}), whereas resveratrol caused >50% inhibition of mycelial colonization of wheat grains at that concentration. However, a concentration of 500 μg mL^{-1} was required for >90% inhibition of growth. In addition, a complete inhibition of production of OTA by *P. verrucosum* was observed with 500 μg mL^{-1} of resveratrol. Only the lowest concentration (50 μg mL^{-1}) was proven to stimulate production of OTA by *A. westerdijkiae*.

Currently, Crespo-Sempere et al. [151] evaluated the oxidative stress and the influence of phenolic antioxidants on production of OTA and growth of *A. carbonarius*. They analyzed the effect of the oxidant menadione and of the antioxidants 3,5-di-tert-butyl-4-hydroxytoluene (BHT), catechin, resveratrol and a polyphenolic extract on growth rate, generation of ROS, production of OTA and gene expression of antioxidant enzymes of *A. carbonarius*. The presence of concentrations of menadione higher than 20 μmol L^{-1} in the medium led to increases in ROS and OTA levels and to a decrease in growth. Exposure to concentrations of BHT ranging from 2.5 to 10 mmol L^{-1} also led to higher levels of ROS and OTA. Other compounds such as catechin and resveratrol did not significantly reduce the generation of ROS, which correlates with an unsignificant reduction of OTA. However, the polyphenol extract showed a statistically significant antioxidant effect when the culture medium was supplemented with 20 and 35 ppm of extract, since the generated ROS were lower as compared to controls.

In contrast to catechin and resveratrol, the polyphenol extract was able to significantly reduce the content of OTA (22%) at the highest concentration tested (35 ppm).

Other studies also showed a reduction in *A. carbonarius* growth in the presence of resveratrol [152]. In addition, Galati et al. [153] observed that some polyphenols such as resveratrol may be metabolized by peroxidase to form pro-oxidant phenoxyl radicals that might generate ROS, H_2O_2 and semiquinone or quinones, all with proven cytotoxicity [154-155].

Crespo-Sempere et al. [151] reported that catechin, resveratrol and the polyphenol extract produced different effects on OTA production levels depending on the concentrations used. As results, they observed a reduction of 22% (35 μmol L^{-1} of polyphenol extract) and an increase of 61% (1,500 μmol L^{-1} of resveratrol), compared to the control treatment. Both the antioxidant and the pro-oxidant capacity of a compound depend on several variables, such as its chemical structure, concentration or mode of action. It should be noted that phenolic antioxidants have been shown to undergo loss of activity and become pro-oxidants at high concentrations. Several studies have been conducted to analyze the effect of resveratrol on production of OTA by *A. carbonarius*, yielding contradictory results. Resveratrol is a phytoalexin produced by grapes in response to fungal invasion. On the other hand, Bavaresco et al. [152] reported that the presence of resveratrol triggers the production of OTA in synthetic must medium. De Rossi et al. [156] found that resveratrol at 0.1 mmol L^{-1} caused a significant inhibition of the production of OTA by *A. carbonarius in vitro*. In contrast, a positive correlation between high levels of OTA and the content of resveratrol in wine was reported by Perrone et al. [157].

Gene Regulation and Antioxidants

Since the biochemical and genetic components of the OTA biosynthetic pathway are not fully understood, and because of the diversity of OTA producers, some recent works have evaluated the molecular regulation of the production of OTA [158-159].

With regards to production of OTA and gene expression, Crespo-Sempere et al. [151] evaluated the effect of low and high concentrations of the antioxidant compounds BHT, catechin and polyphenol extract on the expression of the *sod* and *prx* genes [superoxide dismutase (SOD) enzyme and peroxiredoxin (PRX)]. With BHT, the expression of *sod* and *prx* was down regulated and this down regulation was higher with 10 mmol L^{-1} than with 2.5 mmol L^{-1}. Down regulation was higher in the case of *sod*, and changed from 2.2- to 5.4-fold when the concentration of BHT was increased from 2.5 to 10 mmol L^{-1}. By comparison, the expression of *prx* changed from 2.6- to 3.4-fold. With catechin, an important change in the expression of *sod* and *prx* was observed depending on the concentration used. On the other hand, the expression was not affected at 25 μmol L^{-1}. It was strongly down regulated at 1,500 μmol L^{-1} (10.7- and 8.5-fold for *sod* and *prx*, respectively). In contrast, the expression of *sod* and *prx* was upregulated when culture media were supplemented with a concentration of resveratrol that increased the production of OTA (1,500 μmol L^{-1}). These authors concluded that the above mentioned oxidative enzymes, especially SOD (which was particularly upregulated) play a role in the fungal response to oxidative stress induced by the presence of high concentrations of resveratrol. Finally, the expression of *sod* and *prx* was studied in cultures of *A. carbonarius* supplemented with 20 and 35 ppm of polyphenol extract. Both concentrations led to a reduction in the expression of *sod* and *prx*. However, this down regulation was greater at 20 ppm, in parallel with an increase in the production of OTA. A comparison of the data from production of OTA and gene expression suggests two different patterns in the expression of *sod* and *prx*. Thus, *A. carbonarius* may have two different mechanisms of response to oxidative stress. When resveratrol showed a pro-oxidant behavior that increased production of OTA, the expression of *sod* and *prx* was upregulated. This result suggests that the enzymes SOD and PRX are involved in the cellular response to oxidative stress. In contrast, pro-oxidant concentrations of catechin and the polyphenol extract behaved like menadione. When these compounds had a pro-oxidant effect, which enhanced production of OTA, the expression of genes was down regulated.

CONCLUSION

The overview of the presented data indicates that antioxidants generally suppress production of OTA in several important ochratoxigenic species. The effect of each compound on growth and production of OTA differed among strains and is generally variable, suggesting that species-specific production of OTA and response to phenolic compounds may be influenced by different ecological and developmental factors. Since these substances can either promote or inhibit growth and production of toxin, determining the optimal application doses on natural substrates becomes of great importance. High levels of an antioxidant or antimicrobial compound do not ensure inhibition of the production of a mycotoxin. In fact, they could cause undesirable effects on the organoleptic properties of grains. On the other hand, sub-inhibitory doses together with an inadequate distribution of the preservative, especially at low a_w levels, could stimulate fungal sporulation, growth and secondary metabolism and increase contamination by mycotoxins.

The data suggest that the mechanisms of phenolic antioxidant activity may be directly or indirectly related to the primary metabolism, as evidenced by the effects on fungal growth, or to the secondary metabolism, or to the combination of both.

Further studies will allow to determine if the effects on production of secondary metabolites, such as OTA, are due to an influence on the oxidative stress response of the fungi. In addition, inhibition of OTA by these compounds may be useful to determine differences in gene expression in response to such compounds, and to identify biosynthetic and regulatory genes necessary for production of OTA. Further researches on the bioavailability and fungal dose-response to phenolic antioxidants under ecologically relevant conditions are warranted. Information regarding genetic and physiological responses to antioxidant compounds could lead to targeted intervention strategies to reduce the economic losses caused by contamination with OTA.

REFERENCES

[1] Harris, JP; Mantle, PG. Biosynthesis of ochratoxins by *Aspergillus ochraceus*. *Phytochemistry* 2001, 58, 709-716.

`] World Health Organization (WHO). Total Diet Studies: a recipe for safer food. 2005. Available from: http://www.who.int/foodsafety/chem/TDSrecipe2005enpd (accessed March 2015).

] IARC International Agency for Research on Cancer. Ochratoxin A. *In IARC Monographs on the evaluation of carcinogenic risks to human: some naturally occurring substances; food items and constituents, heterocyclic aromatic amines and mycotoxins*. Geneva, Switzerland, 1993; volume 56, pp. 26-32.

Clark, HA; Snedeker, SM. Ochratoxin A: Its cancer risk and potential for exposure. *Journal of Toxicology Environmental and Health* B 2006, 9, 265–296.

[5] European Commission Commission Regulation EC No 105/2010 amending Regulation (EC) No 1881/2006 setting maximum levels for certain contaminants in foodstuffs as regards ochratoxin A. *Official Journal of the European Union,* 35, 2010, pp. 7-8.

[6] Abarca, ML; Accensi, F; Cano, J; Cabañes, FJ. Taxonomy and significance of black Aspergilli. *Antonie Van Leeuwenhoek* 2004, 86, 33-49.

[7] Bogs, C; Battilani, P; Geisen, R. Development of a molecular detection and differentiation system for ochratoxin A producing *Penicillium* species and its application to analyze the occurrence of *Penicillium nordicum* in cured meats. *International Journal of Food Microbiology* 2006, 107, 39-47.

[8] Battilani, P; Pietri, VA; Giorni, P; Formenti, S; Bertuzzi, T; Toscani, T; Virgili, R; Kozakiewicz, Z. *Penicillium* populations in dry-cured ham manufacturing plants. *Journal of Food Protection* 2007, 70, 975–980.

[9] Kizis, D; Natskoulis, P; Nychas, GJ; Panagou, EZ. Biodiversity and ITS-RFLP characterization of *Aspergillus* section *Nigri* isolates in grapes from four traditional grape-producing areas in Greece. *PLoS One* 2014: 7, 9(4):e93923. doi: 10.1371/journal.pone.0093923. eCollection 2014.

[10] Gil-Serna, J; Patiño, B; Cortes, L; Gonzalez-Jaen, MT; Vazquez, C. *Aspergillus steynii* and *Aspergillus westerdijkiae* as potential risk of OTA contamination in food products in warm climates. *Food Microbiology* 2015, 46, 168-175.

[11] Magan, N; Hope, R; Cairns, V; Aldred, D. Post-harvest fungal ecology: Impact of fungal growth and mycotoxin accumulation in stored grain. *European Journal of Plant Pathology* 2003, 109, 723-730.

[12] Bucheli, P; Taniwaki, MH. Review. Research on the origin, and on the impact of post-harvest handling and manufacturing on the presence of ochratoxin A in coffee. *Food Additives and Contaminants* 2002, 19, 655-665.

[13] Rivera-Carriles, K; Argaiz, A; Palou, E; Lopez-Malo, A. Synergistic inhibitory effect of citral with selected phenolics against *Zygosaccharomyces bailii. Journal of Food Protection* 2005, 68, 602–606.

[14] Ahn, Y; Lee, H; Oh, H; Kim, H; Lee, YH. Antioxidant activity and phenolic composition of citrus peel and seed extracts. *Journal of Agricultural and Food Chemistry* 2005, 46, 2123- 2129.

[15] Etcheverry, M; Torres, A; Ramirez, M; Chulze, S; Magan, N. *In vitro* control of growth and fumonisin production by *Fusarium verticilloides*

and *Fusarium proliferatum* using antioxidants under different water availability and temperature regimes. *Journal of Applied Microbiology* 2002, 92, 624-632.

[16] Reynoso, M; Torres, A; Ramírez, M; Rodríguez, M; Chulze, S; Magan, N. Efficacy of antioxidant mixtures on growth, fumonisins production and hydrolytic enzyme production by *Fusarium verticillioides* and *F. proliferatum in vitro* on maize-based media. *Mycological Research* 2002, 106, 1093-1099.

[17] Torres, A; Ramírez, M; Arroyo, M; Chulze, S; Magan, N. Potencial use of antioxidants for control of growth and fumonisin production by *Fusarium proliferatum* and *Fusarium verticilloides* on whole maize grain. *International Journal of Food Microbiology* 2003, 83, 319-324.

[18] Selvi, A; Joseph, G; Jayaprakasha, G. Inhibition of growth and aflatoxin production in *Aspergillus flavus* by *Garcinia indica* extract and its antioxidant activity. *Food Microbiology* 2003, 20, 455–460.

[19] Nesci, A; Rodriguez, M; Etcheverry, M. Control of *Aspergillus* growth and aflatoxins production using antioxidants at different conditions of water activity and pH. *Journal of Applied Microbiology* 2003, 95, 279-287.

[20] Joseph, G; Jayaprakasha, G; Selvi, AT; Jena, B; Sakariah, KK. Antiaflatoxigenic and antioxidant activities of *Garcinia* extracts. *International Journal of Food Microbiology* 2005, 101, 153–160.

[21] Farnochi, C; Torres, A; Magan, N; Chulze, S. Effect of antioxidants and competing mycoflora on *Fusarium verticilloides* and *F. proliferatum* populations and fumonisin production on maize grain. *Journal of Stored Research* 2005, 41, 211-219.

[22] Passone, M; Resnik S; Etcheverry M. *In vitro* effect of phenolic antioxidants on germination, growth and aflatoxin B_1 accumulation by peanut *Aspergillus* section *Flavi*. *Journal of Applied Microbiology* 2005, 99, 682-691.

[23] Passone, M; Resnik S; Etcheverry M. Antiaflatoxigenic property of food grade antioxidants under different conditions of water activity in peanut grains. *International Journal of Food Microbiology* 2007, 118, 8-14.

[24] Palumbo, JD; O'Keeffe, TL; Mahoney, NE. Inhibition of ochratoxin A production and growth of *Aspergillus* species by phenolic antioxidant compounds. *Mycopathologia* 2007, 164, 241-248.

[25] Bisogno, F; Mascoti, L; Sanchez, C; Garibotto, F; Giannini, F; Kurina-Sanz, M; Enriz, R. Structure-antifungal activity relationship of cinnamic

acid derivatives. *Journal of Agricultural and Food Chemistry* 2007, 55, 10635-10640.

[26] Romero, S; Alberto, M; Vaamonde, G. Effect of gallic acid on *Aspergillus carbonarius* growth and ochratoxin A production. *World Mycotoxin Journal* 2010, 3, 45-48.

[27] RASFF. Annual Report 2011. Publications Office of the European Union, Luxemburg, 2012.

[28] Azpilicueta, CA; Arbeloa, MI; de Maquirriain, PFJ. Natural and synthetic occurring forms of the ochratoxins. In: Papadopoulos KN. (Ed.) *Food chemistry research developments.* New York: Nova Science Publishers, 2008, 116.

[29] Ringot, D; Chango, A; Schneider, YJ; Larondelle, Y. Toxicokinetics and toxicodynamics of ochratoxin A, an update. *Chemico-Biological Interactions* 2006, 19, 18–46.

[30] Pfohl-Leszkowicz, A; Manderville, RA. Ochratoxin A: an overview on toxicity and carcinogenicity in animals and humans. *Molecular Nutrition and Food Research* 2007, 51, 61-99.

[31] Pfohl-Leszkowicz, A; Manderville, RA. An update on direct genotoxicity as a molecular mechanism of ochratoxin A carcinogenicity. *Chemical Research in Toxicology* 2012, 25, 252-262.

[32] Mantle, PG; Faucet-Marquis, V; Manderville, RA; Squillaci, B; Pfohl-Leszkowicz, A. Structures of covalent adducts between DNA and ochratoxin A: a new factor in debate about genotoxicity and human risk assessment. *Chemical Research in Toxicology* 2010, 23, 89-98.

[33] Bozic, Z.; Duancic, V.; Kraus, O.; Skljarov, I. Balkan endemic nephropathy: Still a mysterious disease. *European Journal of Epidemiology* 1995, 11, 235–238.

[34] Castegnaro, M; Bartsch, H; Chernozemsky, I. Endemic nephropathy and urinary tract tumors in the Balkans. *Cancer Research* 1987, 47, 3608–3609.

[35] Ringot, D; Chango, A. Risk assessment of ochratoxin A (OTA). In Rai M, Varma A. (Eds.). *Mycotoxins in food, feed and bioweapons.* Dordrecht, London, New York: Springer, 2010, Chapter 18, pp. 307-328.

[36] Van Der Merwe, KJ; Steyn, PS; Fourie, L. Ochratoxin A, a toxic metabolite produced by *Aspergillus ochraceus*. *Wilh Nature* 1965, 205, 1112–1113.

[37] Frisvad, JC; Frank, JM; Houbraken, JAMP, Kuijpers, AFA; Samson, RA. New ochratoxin A producing species of *Aspergillus* section *Circumdati*. *Studies in Mycology* 2004, 50, 23-43.

[38] Cabañes, FJ; Bragulat, MR; Castellá, G. Ochratoxin A producing species in the genus *Penicillium*. *Toxins* 2010, 2, 1111-1120.

[39] Perrone, G; De Girolamo, A; Sarigiannis, Y; Haidukowski, ME; Visconti, A. Occurrence of ochratoxin A, fumonisin B_2 and black Aspergilli in raisins from Western Greece regions in relation to environmental and geographical factors. *Food Additives and Contaminants Part A* 2013, 30, 1339-1347.

[40] Pitt, JI; Hocking, AD. Fungi and food spoilage, (3ed edition). Sidney, *Australia, CSIRO Division of Food Science and Technology,* Sydney Academic, Press. 2009.

[41] Palumbo, JD; O'Keeffe TL; Vasquez SJ; Mahoney NE. Isolation and identification of ochratoxin A-producing *Aspergillus* section *Nigri* strains from California raisins. *Letters in Applied Microbiology* 2011, 52, 330-336.

[42] Magnoli, C; Astoreca, A; Ponsone, L; Combina, M; Palacio, G; Rosa, CAR; Dalcero, A. Survey of mycoflora and ochratoxin A in dried vine fruits from Argentina markets. *Letters in Applied Microbiology* 2004, 39, 326–331.

[43] Ponsone, ML; Combina, M; Dalcero, A; Chulze, S. Ochratoxin A and ochratoxigenic *Aspergillus* species in Argentinean wine grapes cultivated under organic and non-organic systems. *International Journal of Food Microbiology* 2007, 114, 131-135.

[44] Ponsone, ML; Chiotta ML; Combina, M; Dalcero, AM; Chulze, SN. Fate of ochratoxin A content in Argentinian red wine during a pilot scale vinification. *Revista Argentina de Microbiología* 2009, 41, 245–250.

[45] Chiotta, ML; Ponsone, ML; Sosa, DM; Combina, M; Chulze, SN. Biodiversity of *Aspergillus* section *Nigri* populations in Argentinian vineyards and ochratoxin A contamination. *Food Microbiology* 2013, 36, 182-190.

[46] Joosten, HM; Goetz, J; Pittet, A; Schellenberg, M; Bucheli, P. Production of ochratoxin A by *Aspergillus carbonarius* on coffee cherries. *International Journal of Food Microbiology* 2001, 65, 39-44.

[47] Bonvehi, JS. Occurrence of ochratoxin A in cocoa products and chocolate. *Journal of Agriculture and Food Chemistry* 2004, 52, 6347–6352.

[48] Aoyama, K; Nakajima, M; Tabata, S; Ishikuro, E; Tanaka, T; Norizuki, H; Itoh, Y; Fujita, K; Kai, S; Tsutsumi, T; Takahashi, M; Tanaka, H; Iizuka, S; Ogiso, M; Maeda, M; Yamaguchi, S; Sugiyama, K; Sugita-Konishi, Y; Kumagai, S. Four-year surveillance for ochratoxin A and fumonisins in retail foods in Japan. *Journal of Food Protection* 2010, 73, 344-352.

[49] Ratola, N; Barros, P; Simoes, T; Cerdeira, A; Venancio, A; Alves, A. Worldwide interlaboratory study on the determination of ochratoxin A in different wine type samples. *Talanta* 2006, 70, 720–731.

[50] Tessini, C; Mardones, C; von Baer, D; Vega, M; Herlitz, E; Saelzer, R; Silva, J; Torres, O. Alternatives for sample pre-treatment and HPLC determination of ochratoxin A in red wine using fluorescence detection. *Analytica Chimica Acta* 2010, 660, 119-126.

[51] Medina, A; Jiménez, M; Gimeno-Adelantado, J; Valle-Algarra, F; Mateo, R. Determination of ochratoxin A in beer marketed in Spain by liquid chromatography with fluorescence detection using lead hydroxyacetate as a clean-up agent. *Journal of Chromatography* A 2005, 1083, 7–13.

[52] Reinsch, M; Topfer, A; Lehmann, A; Nehls, I; Panne, U. Determination of ochratoxin A in beer by LC-MS/MS ion trap detection. *Food Chemistry* 2007, 100, 312-317.

[53] Bellver Soto, J; Fernández-Franzón, M; Ruiz, MJ; Juan-García, A. Presence of ochratoxin A (OTA) mycotoxin in alcoholic drinks from southern European countries: wine and beer. *Journal of Agriculture and Food Chemistry* 2014, 62, 7643-7651.

[54] Battilani, P; Magan, N; Logrieco, A. European research on ochratoxin A in grapes and wine. *International Journal of Food Microbiology* 2006, 111, S2–S4.

[55] Varga, J; Kozakiewicz, Z. Ochratoxin A in grapes and grape derived products. *Trends in Food Science and Technology* 2006, 17, 72–81.

[56] Lasram, S; Bellí, N; Chebil, S; Nahla, Z; Ahmed, M; Sanchís, V; Ghorbel, A. Short communication: Occurrence of ochratoxigenic fungi and ochratoxin A in grapes from a Tunisian vineyard. *International Journal of Food Microbiology* 2007, 114, 376–379.

[57] García-Cela, E; Crespo-Sempere, A; Gil-Serna, J; Porqueres, A; Marin, S. Fungal diversity, incidence and mycotoxin contamination in grapes from two agro-climatic Spanish regions with emphasis on *Aspergillus* species. *Journal of the Science of Food and Agriculture* 2015, 95, 1716-1729.

[58] Akdeniz, AS; Ozden, S; Alpertunga, B. Ochratoxin A in dried grapes and grape-derived products in Turkey. *Food Additives and Contaminants Part B* 2013, 6, 265-269.

[59] Bircan, C. Incidence of ochratoxin A in dried fruits and co-occurrence with aflatoxins in dried figs. *Food and Chemical Toxicology* 2009, 47, 1996–2001.

[60] Karbancıoglu-Guler, F; Heperkan, D. Natural occurrence of ochratoxin A in dried figs. *Analytica Chimica Acta* 2008, 617, 32–36.

[61] Magnoli, C; Astoreca, A; Ponsone, ML; Fernández-Juri, MG; Barberis, C; Dalcero, A. Ochratoxin A and *Aspergillus* section *Nigri* in peanut seeds at different months of storage in Córdoba, Argentina. *International Journal of Food Microbiology* 2007a, 119, 213–218.

[62] Magnoli, C; Astoreca, A; Chiacchiera, SM; Dalcero, A. Occurrence of ochratoxin A and ochratoxigenic mycoflora in corn and corn based foods and feeds in some South American countries. *Mycopathologia* 2007b, 163, 249–260.

[63] Duarte, SC; Pena, A; Lino, CM. A review on ochratoxin A occurrence and effects of processing of cereal and cereal derived food products. *Food Microbiology* 2010, 27, 187–198.

[64] Juan, C; Pena, A; Lino, C; Moltó, JC; Mañes, J. Levels of ochratoxin A in wheat and maize bread from the central zone of Portugal. *International Journal of Food Microbiology* 2008, 127, 284-289.

[65] Zhihong, L; Kunlun, H; Yunbo, L. Ochratoxin A and ochratoxin-producing fungi on cereal grain in China: a review. *Food Additives and Contaminants Part A* 2015, 32, 461-470.

[66] Coronel, MB; Marín, S; Cano-Sancho, G; Ramos, AJ; Sanchis, V. Exposure assessment to ochratoxin A in Catalonia (Spain) based on the consumption of cereals, nuts, coffee, wine, and beer. *Food Additives and Contaminants Part A* 2012, 29, 979-993.

[67] Matrella, R.; Monaci, L.; Milillo, M.A.; Palmesano, F.; Tantillo, M.G. Ochratoxin A determination in paired kidneys and muscle samples from swines slaughtered in southern Italy. *Food Control* 2006, 17, 114–117.

[68] Aoudia, N; Tangni, EK; Larondelle, Y. Distribution of ochratoxin A in plasma and tissues of rats fed a naturally contaminated diet amended with micronized wheat fibers: Effectiveness of mycotoxin sequestering activity. *Food and Chemistry Toxicology* 2008, 46, 871–878.

[69] Coronel, MB; Sanchis, V; Ramos, AJ; Marin, S. Assessment of the exposure to ochratoxin A in the province of Lleida, Spain. *Food Chemistry and Toxicology* 2009, 47, 2847-2852.

[70] di Giuseppe, R; Bertuzzi, T; Rossi, F; Rastelli, S; Mulazzi, A; Capraro, J; de Curtis, A; Iacoviello, L; Pietri, A. Plasma ochratoxin A levels, food consumption, and risk biomarkers of a representative sample of men and women from the Molise region in Italy. *European Journal of Nutrition* 2012, 51, 851-860.

[71] Bullerman, LB; Bianchini, A. Stability of mycotoxins during food processing. *International Journal of Food Microbiology* 2007, 119, 140–146.

[72] Puntaric, D; Bosnir, J; Smit, Z; Baklaic, Z. Ochratoxin A in corn and wheat: geographical association with endemic nephropathy. *Croatian Medical Journal* 2001, 42, 175–180.

[73] Pardo, E; Marín, S; Sanchís, V; Ramos, AJ. Prediction of fungal growth and ochratoxin A production by *Aspergillus ochraceus* on irradiated barley grain as influenced by temperature and water activity. *International Journal of Food Microbiology* 2004, 95, 79–88.

[74] Pardo, E; Marín, S; Sanchís, V; Ramos, AJ. Impact of relative humidity and temperature on visible fungal growth and OTA production of ochratoxigenic *Aspergillus ochraceus* isolates on grapes. *Food Microbiology* 2005a, 22, 383–389.

[75] Pardo, E; Ramos, AJ; Sanchís, V; Marín, S. Modeling of the effects of water activity and growth of ochratoxigenic isolates of *Aspergillus ochraceus* on a green coffee-based medium. *International Journal of Food Microbiology* 2005b, 98, 1–9.

[76] Pardo, E; Marín, S; Ramos, AJ; Sanchís, V. Ecophysiology of ochratoxigenic *Aspergillus ochraceus* and *Penicillium verrucosum* isolates. Predictive models for fungal spoilage prevention-a review. *Food Additives and Contaminants* 2006, 23, 398–410.

[77] Arroyo, M; Aldred, D; Magan, N. Environmental factors and weak organic acid interactions have differential effects on control of growth and ochratoxin A production by *Penicillium verrucosum* isolates in bread. *International Journal of Food Microbiology* 2005, 98, 223-231.

[78] Czaban, J; Wróblewska, B; Stochmal, A; Janda, B. Growth of *Penicillium verrucosum* and production of ochratoxin A on nonsterilized wheat grain incubated at different temperatures and water content. *Polish Journal of Microbiology* 2006, 55, 321–331.

[79] Aldred, D; Cairns-Fuller, V; Magan, N. Environmental factors affect efficacy of some essential oils and resveratrol to control growth and ochratoxin A production by *Penicillium verrucosum* and *Aspergillus*

westerdijkiae on wheat grain. *Journal of Stored Products Research* 2008, 44, 341–346.

[80] Kapetanakou, AE; Panagou, EZ; Gialitaki, M; Drosinos, EH; Skandamis, PN. Evaluating the combined effect of water activity, pH and temperature on ochratoxin A production by *Aspergillus ochraceus* and *Aspergillus carbonarius* on culture medium and Corinth raisins. *Food Control* 2009, 20, 725–732.

[81] Muñoz, K; Vega, M; Rios, G; Geisen, R; Degen, GH. Mycotoxin production by different ochratoxigenic *Aspergillus* and *Penicillium* species on coffee- and wheat-based media. *Mycotoxin Research* 2011, 27, 239-247.

[82] Astoreca, A; Magnoli, C; Barberis, C; Combina, M; Dalcero, A. Growth and ochratoxin A production by *Aspergillus niger* group strains in coffee beans in relation to environmental factors. *World Mycotoxin Journal* 2010a, 3, 59–65.

[83] Astoreca, A; Magnoli C; Dalcero, A. Ecophysiology of *Aspergillus* section *Nigri* species potential ochratoxin A producers. *Toxins* 2010b, 2, 2593-2605.

[84] Astoreca, A; Magnoli C; Barberis, C; Combina, M; Dalcero, A. *Aspergillus carbonarius* growth and ochratoxin A production in irradiated dried grapes under different water activity and temperature conditions. *World Mycotoxin Journal* 2010c, 3, 175–182.

[85] Oueslati, S; Lasram, S; Ramos, AJ; Marin, S; Mliki, A; Sanchís, V; Ghorbel, A. Alternating temperatures and photoperiod effects on fungal growth and ochratoxin A production by *Aspergillus carbonarius* isolated from Tunisian grapes. *International Journal of Food Microbiology* 2010, 139, 210–213.

[86] Magan, N; Aldred, D; Hope, R; Mitchell, D. Environmental factors and interactions with mycobiota of grain and grapes: effects on growth, deoxynivalenol and ochratoxin production by *Fusarium culmorum* and *Aspergillus carbonarius*. *Toxins (Basel)* 2010, 2, 353-266.

[87] Alborch, L; Bragulat, MR; Abarca, ML; Cabañes FJ. Effect of water activity, temperature and incubation time on growth and ochratoxin A production by *Aspergillus niger* and *Aspergillus carbonarius* on maize kernels. *International Journal of Food Microbiology* 2011, 147, 53-57.

[88] Passamani, FR; Hernandes, T; Lopes, NA; Bastos, SC; Santiago, WD; Cardoso, MD; Batista, LR. Effect of temperature, water activity, and pH on growth and production of ochratoxin A by *Aspergillus niger* and

Aspergillus carbonarius from Brazilian grapes. *Journal of Food Protection* 2014, 77, 1947-1952.

[89] Belli, N; Pardo, E; Marin, S; Farre, G; Ramos, AJ; Sanchis, V. Occurrence of ochratoxin A and toxigenic potential of fungal isolates from Spanish grapes. *Journal of the Science of Food and Agriculture* 2004, 84, 541–546.

[90] Centre for Food Safety. Ochratoxin A in food. Food and Environmental Hygiene Department. Risk Assessment Studies Report no. 23. *The Government of the Hong Kong Special Administrative Region,* Hong Kong, 2006.

[91] Atalla, MM; Hassanein, NM; El-Beih, AA; Youssef, YAG. Mycotoxin production in wheat grains by different Aspergilli in relation to different relative humidities and storage periods. *Nahrung* 2003, 47, 6–10.

[92] Gimeno, A. *Micotoxinas, introducción. XVII Congreso centroamericano y del Caribe de avicultura,* 2000. Available from: http://www.engormix.com (Accessed: October 2008).

[93] Magan, N; Aldred, D. Conditions of formation of ochratoxin A in drying, transport and in different commodities. *Food Additives and Contaminants* 2005, 1, 10-16.

[94] Codex Alimentarius Commission. *Codex Alimentarius Commission Alinorm 03/12 Appendix XII. Technical Report for the 34th Session of the Codex committee on Food Additives and Contaminants*; Codex Alimentarius Commission: Rome, Italy, 2003; pp. 100–104.

[95] Belli, N; Marin S; Argilés, E; Ramos, AJ; Sanchis, V. Effect of chemical treatments on ochratoxigenic fungi and common mycobiota of grapes (*Vitis vinifera*). *Journal of Food Protection* 2007, 70, 157–163.

[96] Medina, A; Jiménez, M; Mateo, R; Magan, N. Efficacy of natamycin for control of growth and ochratoxin A production by *Aspergillus carbonarius* strains under different environmental conditions. *Journal of Applied Microbiology* 2007, 103, 2234–2239.

[97] Cozzi, G; Somma, S; Haidukowski, M; Logrieco, AF. Ochratoxin A management in vineyards by *Lobesia botrana* biocontrol. *Toxins (Basel)* 2013, 2, 49-59.

[98] Cubaiu, L; Abbas, H; Dobson, AD; Budroni, M; Migheli, Q. A *Saccharomyces cerevisiae* wine strain inhibits growth and decreases ochratoxin A biosynthesis by *Aspergillus carbonarius* and *Aspergillus ochraceus. Toxins (Basel)* 2012, 4, 1468-1481.

[99] Magan, N; Aldred, D. Post-harvest control strategies; mycotoxins from the field to the table: Minimizing mycotoxins in the food chain. *International Journal of Food Microbiology* 2007a, 119, 131-139.

[100] Magan, N; Aldred, D. Why do fungi produce mycotoxins? In Dijksterhuis J; Samson R.A. (Eds.), Food mycology: a multifaceted approach to fungi and food. Boca Raton FL, USA: Taylor & Francis. 2007b, pp. 121–133.

[101] Magan, N; Aldred, D. Environmental fluxes and fungal interactions: maintaining a competitive edge. In van West P; Avery S; Stratford M (Eds.*), Stress in yeasts and filamentous fungi*, Amsterdam, The Netherlands: Elsevier Ltd. 2007c, pp. 19–35.

[102] Amézqueta, S; González-Peñas, E; Lizarraga, T; Murillo-Arbizu, M; López de Cerain, A. A simple chemical method reduces ochratoxin A in contaminated cocoa shells. *Journal of Food Protection* 2008, 71, 1422-1426.

[103] Aroyeun, SO; Adegoke, GO. Reduction of ochratoxin A (OTA) in spiked cocoa powder and beverage using aqueous extracts and essential oils of *Agramomum danielli. African Journal of Biotechnology* 2007, 6, 612-616.

[104] Hua, H; Xing, F; Selvaraj, JN; Wang, Y; Zhao, Y; Zhou, L; Liu, X; Liu Y. Inhibitory effect of essential oils on *Aspergillus ochraceus* growth and ochratoxin A production. *PLoS One.* 2014, 9(9):e108285. doi: 10.1371/journal.pone.0108285. eCollection 2014.

[105] Ponsone, ML; Chiotta, ML; Palazzini, JM; Combina, M; Chulze S. Control of ochratoxin A production in grapes. *Toxins (Basel)* 2012, 364-372.

[106] Ersilia, A; Poiana, MA; Sumalan, RM. Mycoflora and ochratoxin A control in wheat grain using natural extracts obtained from wine industry by-products. *International Journal of Molecular Sciences* 2012, 13, 4949–4967.

[107] Halliwell, B; Aeschbach, R; Loliger, J; Aruoma O. The characterization of antioxidants. *Food Chemistry and Toxicology* 1995, 33, 601–617.

[108] Halliwell, B. Food derived antioxidants. Evaluating their importance in food and *in vivo*. *Food Science and Agricultural Chemistry* 1999, 1, 67–109.

[109] Warner, C; Brumley, W; Daniels, D; Joe, F; Fazio, T. Reactions of antioxidants in foods. *Food Chemistry and Toxicology* 1986a, 24, 1015–1019.

[110] Lam L; Pai, R; Wattenberg, L. Synthesis and chemical carcinogen inhibitory activity of 2-tert-butyl-4-hydroxyanisole. *Journal of Medicinal Chemistry* 1979, 22, 569–571.

[111] Page, B; Charbonneau, C. Liquid chromatographic determination of seven antioxidants in dry food. *Journal of the Association of Official Analytical Chemists* 1989, 72, 259–265.

[112] Soni, M; Carabin, I; Burdock, G. Safety assessment of esters of phydroxybenzoic acid (parabens). *Food Chemistry and Toxicology* 2005, 43, 985–1015.

[113] Soni, M; Burdock, G; Taylor, S; Greenberg, N. Safety assessment of propyl paraben: a review of the published literature. *Food Chemistry and Toxicology* 2001, 39, 513-532.

[114] Codex alimentarius, (2006). Food additive details. *Update up to the twenty-ninth session of the codex alimentarius commission.* The Joint FAO/WHO Committee on Food Additives. Available in: www.codexalimentarius.net/web/jecfa (Accessed June 2015).

[115] Farah, A; Donangelo, CM. Penolic compounds in coffee. *Brazilian Journal of Plant Physiology* 2006, 18, 23–36.

[116] Kanner, J; Frankel, E; Granit, R; German, B; Kinsella, JE. Natural antioxidants in grapes and wines. *Journal of Agricultural and Food Chemistry* 1994, 42, 64–69.

[117] Etoh, H; Murakami, K; Yogoh, T; Ishikawa, H; Fukuyama, Y; Tanaka, H. Anti-oxidative compounds in barley tea. *Bioscience Biotechnology and Biochemistry* 2004, 68, 2616–2618.

[118] Bonoli, M; Verardo, V; Marconi, E; Caboni, MF. Antioxidant phenols in barley (*Hordeum vulgare* L.) flour: comparative spectrophotometric study among extraction methods of free and bound phenolic compounds. *Journal of Agriculture and Food Chemistry* 2004, 52, 5195–5200.

[119] Mahoney, N; Molyneux, RJ. Phytochemical inhibition of aflatoxigenicity in *Aspergillus flavus* by constituents of walnut (*Juglans regia*). *Journal of Agricultural and Food Chemistry* 2004, 52, 1882–1889.

[120] Sang, S; Lapsley, K; Jeong, WS; Lachance, PA; Ho, CT; Rosen, RT. Antioxidative phenolic compounds isolated from almond skins (*Prunus amygdalus Batsch*). *Journal of Agricultural and Food Chemistry* 2002, 50, 2459–2463.

[121] Takeoka, GR; Dao, LT. Antioxidant constituents of almond [*Prunus dulcis* (Mill.) D.A. Webb] hulls. *Journal of Agricultural and Food Chemistry* 2003, 51, 496–501.

[122] Goli, AH; Barzegar, M; Sahari, MA. Antioxidant activity and total phenolic compounds of pistachio (*Pistachia vera*) hull extracts. *Food Chemistry* 2004, 92, 521–255.

[123] Seeram, NP; Zhang, Y; Henning, SM; Lee, R; Niu, Y; Lin, G; Heber, D. Pistachio skin phenolics are destroyed by bleaching resulting in reduced antioxidative capacities. *Journal of Agricultural and Food Chemistry* 2006, 54, 7036–7040.

[124] Barberis, C; Astoreca, A; Asili, R; Fernández-Juri, M; Chulze, S; Magnoli, C; Dalcero, A. *In vitro* control of growth and ochratoxin A production by butylatedhydroxyanisole in *Aspergillus* section *Nigri* species. *Food Control* 2009a, 20, 709-715.

[125] Barberis, C; Astoreca, A; Fernández-Juri, M; Chulze, S; Magnoli, C; Dalcero, A. Use of propyl paraben to control growth and ochratoxin A production by *Aspergillus* section *Nigri* species on peanut meal extract agar. *International Journal of Food Microbiology* 2009b, 136, 133-136.

[126] Marín, S; Sanchis, V; Viñas, I; Canela, R; Magan, N. Effect of water activity and temperature on growth and fuminisin B_1 and B_2 production by *Fusarium proliferatum* and *F. moniliforme* on maize grain. *Letters in Applied Microbiology* 1995, 21, 298-301.

[127] Marín, S; Sanchis, V; Sanz, D; Castel, I; Ramos, A; Canela, R; Magan, N. Control of growth and fumonisin B_1 production by *Fusarium verticilloides* and *Fusarium proliferatum* isolates in moist maize with propionate preservatives. *Food Additives and Contaminants* 1999, 16, 12, 555-563.

[128] Thompson, D. Effect of buthylated hydroxyanisole on conidial germination of toxigenic species of *Aspergillus flavus* and *A. parasiticus*. *Mycologia* 1991, 54, 375-377.

[129] Thompson, D. Inhibition of micelial growth of mycotoxigenic fungi by phenolic antioxidants. *Mycologia* 1992, 54, 791-793.

[130] Thompson, D. Minimum inhibitory concentrations of esters of *p*-hydroxybenzoic acid (paraben) combinations against toxigenic fungi. *Journal of Food Protection* 1994, 57, 133-135.

[131] Marin, S; Guynot, M; Neira, P; Bernadó, M; Sanchis, V; Ramos, A. Risk assessment of the use of sub-optimal levels of weak-acid preservatives in the control of mould growth on bakery products. *International Journal of Food Microbiology* 2002, 79, 203-211.

[132] Romero, SM; Alberto, MR; Manca de Nadra, MC; Vaamonde, G. Inhibition of growth and ochratoxin A biosynthesis in *Aspergillus*

carbonarius by flavonoid and nonflavonoid compounds. *Mycotoxin Research* 2009, 25, 165-170.
[133] Barberis, C; Astoreca, A; Fernandez-Juri, MG; Dalcero, A; Magnoli, C. Effect of antioxidant mixtures on growth and ochratoxin A production of *Aspergillus* section *Nigri* species under different water activity conditions on peanut meal extract agar. *Toxins* 2010a, 2, 1399-1413.
[134] Nesci, A; Gsponer, N; Etcheverry, M. Natural maize phenolic acids for control of aflatoxigenic fungi on maize. *Journal of Food Science* 2007, 72, 180-185.
[135] Smith, J; Moss, M. *Mycotoxins. Formation, analyses and significance.* Wiley, Chichester, 1985; pp. 10–31.
[136] Lacey, J. Prevention of mould growth and mycotoxin production through control of environmental factors. In Natori, S; Hashimoto, K; Ueno, Y. (Eds.), *Mycotoxins and phycotoxins, bioactive molecules.* Elsevier Science, Amsterdam, 1989, Volume 10, pp. 161–169.
[137] Liewen, M. *Antifungal food additives. In* Arora, D; Mukerji, K; Marth, E. (Eds.) Marcell dekker, New York, 1991, Volume 3, pp. 541-552.
[138] Barberis, C; Astoreca, A; Dalcero, A; Magnoli, C. Food-grade antioxidants and antimicrobials to control growth and ochratoxin A production by *Aspergillus* section *Nigri* on peanut kernels. *Journal of Food Protection* 2010b, 73, 1493-1501.
[139] Keshri, G; Voysey, P; Magan, N. Early detection of spoilage moulds production patterns and quantitative. *Journal of Applied Microbiology* 2002, 92, 165–172.
[140] Marin, S; Sanchis, V; Magan, N. Effect of water activity on hydrolytic enzyme production by *Fusarium moniliforme* and *Fusarium proliferatum* during colonization of maize. *International Journal of Food Microbiology* 1998, 42, 185–194.
[141] Alam, S; Shah, HU; Magan, N. Water availability affects extracellular hydrolytic enzyme production by *Aspergillus flavus* and *Aspergillus parasiticus*. *World Mycotoxin Journal* 2009, 2, 313–224.
[142] Sohail, M; Naseeb, S; Sherwani, SK; Sultana, S; Aftab, S; Shahzad, S; Ahmad A, Khan SA. Distribution of hydrolytic enzymes among native fungi: *Aspergillus* the pre-dominant genus of hydrolase producer. *Pakistan Journal of Botany* 2009, 41, 2567–2582.
[143] Barberis, C; Landa, F; Dalcero, AM; Magnoli, C. Effect of butylatedhydroxyanisole and propyl paraben on hydrolytic enzyme production by ochratoxigenic *Aspergillus* section *Nigri* strains isolated

from peanut grains. *Revista Iberoamericana de Micología* 2014, 31, 131-136.

[144] Degré, R; Sylvestre, M. Effect of butylated hydroxyanisole on the cytoplasmic membrane of *Staphylococcus aureus* Wood. *Journal of Food Protection* 1983, 46, 206-209.

[145] Celimene, C; Micales, J; Ferge, L; Young, R. Efficacy of pinosylvins against white-rot and brown-rot fungi. *Holzforschung* 1999, 53, 491-497.

[146] Jayashree, T; Subramanyam, C. Oxidative stress as a prerequisite for aflatoxin production by *Aspergillus parasiticus. Free Radical Biology and Medicine* 2000, 29, 981–985.

[147] Ponts, N; Pinson-Gadais, L; Verdal-Bonnin, MN; Barreau, C; Richard-Forget, F. Accumulation of deoxynivalenol and its 15-acetylated form is significantly modulated by oxidative stress in liquid cultures of *Fusarium graminearum. FEMS Microbiology Ecology* 2006, 258, 102–107.

[148] Narasaiah, K; Sashidhar, R; Subramanyam, C. Biochemical analysis of oxidative stress in the production of aflatoxin and its precursor intermediates. *Mycopathologia* 2006, 162, 179-189.

[149] Kim, JH; Yu, J; Mahoney, N; Chan, KL; Molyneux, RJ; Varga, J; Bhatnagar, D; Cleveland, TE; Nierman, WC; Campbell, BC; Elucidation of the functional genomics of antioxidant-based inhibition of aflatoxin biosynthesis. *International Journal of Food Microbiology* 2008, 122, 49–60.

[150] Fanelli, C; Taddei, F; Nisini, PT; Jestoi, M; Ricelli, A; Visconti, A; Fabbri, AA. Use of resveratrol and BHA to control growth and mycotoxin production in wheat and maize seeds. *Aspects of Applied Biology* 2003, 68, 63-72.

[151] Crespo-Sempere, A; Selma-Lázaro, C; Palumbo, JD; González-Candelas, L; Martínez-Culebras, PV. Effect of oxidant stressors and phenolic antioxidants on the ochratoxigenic fungus *Aspergillus carbonarius. Journal of the Science of Food and Agriculture* 2015, doi: 10.1002/jsfa.7077.

[152] Bavaresco, L; Vezzulli, S; Battilani, P; Giorni, P; Pietri, A; Bertuzzi, T. Effect of ochratoxin A-producing Aspergilli on stilbenic phytoalexin synthesis in grapes. *Journal of Agriculture and Food Chemistry* 2003, 51, 6151–6157.

[153] Galati, G; Sabzevari, O; Wilson, JX; O'Brien, PJ. Prooxidant activity and cellular effects of the phenoxyl radicals of dietary flavonoids and other polyphenolics. *Toxicology* 2002, 177, 91–104

[154] Lambert, JD; Sang, S; Yang, CS. Possible controversy over dietary polyphenols: benefits vs risks. *Chemical Research in Toxicology* 2007, 20, 583–585.

[155] Sang, S; Yang, I; Buckley, B; Ho, CT; Yang, CS. Autoxidative quinone formation *in vitro* and metabolite formation *in vivo* from tea polyphenol (−)-epigallocatechin-3-gallate: studied by real-time mass spectrometry combined with tandem mass ion mapping. *Free Radical Biology and Medicine* 2007, 43, 362–371.

[156] De Rossi, P; Ricelli, A; Reverberi, M; Bello, C; Fabbri, AA; Fanelli, C; De Rossi, A; Corradini, D; Nicoletti, I. Grape variety related trans-resveratrol induction affects *Aspergillus carbonarius* growth and ochratoxin A biosynthesis. *International Journal of Food Microbiology* 2012, 156, 127–132.

[157] Perrone, G; Nicoletti, I; Pascale, M; De Rossi, A; De Girolamo, A; Visconti, A. Positive correlation between high levels of ochratoxin A and resveratrol-related compounds in red wines. *Journal of Agricultural and Food Chemistry* 2007, 55, 6807–6812.

[158] Geisen, R. Molecular monitoring of environmental conditions influencing the induction of ochratoxin A biosynthesis genes in *Penicillium nordicum. Molecular Nutrition and Food Research* 2004, 48, 532–540.

[159] O'Callaghan, J; Stapleton, PC; Dobson, AD. Ochratoxin A biosynthetic genes in *Aspergillus ochraceus* are differentially regulated by pH and nutritional stimuli. *Fungal Genetics and Biology* 2006, 43, 213–421.

In: Ochratoxins
Editor: Daniel Porter

ISBN: 978-1-63483-693-7

Chapter 5

ALTERNATIVE METHOD BY NATURAL PRODUCTS TO OCHRATOXIN CONTROL

Joana D'arc Felicio and Edlayne Gonçalez*

Centrode Pesquisa e Desenvolvimento de Sanidade Animal,
Instituto Biologico, São Paulo, Brazil

ABSTRACT

Ochratoxins are mycotoxins produced principally by *Aspergillus* and *Penicillium* species, and they are found in commodities such as cereals coffee, grape and derived products of them. Ochratoxins exhibit nephrotoxic, immunosuppressive, teratogenic and carcinogenic proprieties, as well as Balkan Endemic Nephropathy have been attributed to this mycotoxin. The ochratoxin A (OTA) is the most potent of them.OTA removal from foods and feedstuffs is not feasible, so prevention management is the best method for controlling the contamination. However, in the case of the contamination occur, a variety of approaches including physical, biological andchemical, treatments to minimize the risk, decontaminate or eliminate the fungi have been attempted. A way to prevent or interfere with fungal growth and ochratoxins production is by use of natural antifungal agents. The use of natural compounds, like essential oils, with lower toxicity of the synthetic products can be a good alternative to the ochratoxin control. This chapter will focus on the use of plants extracts, essential oils and substances

* Corresponding author: Telephone: +55-11-50871753; fax: +55-11-50871754. E-mail address: goncalez@biologico.sp.gov.br (E. Gonçalez).

isolated from them for the control of ochratoxigenic fungi and ochratoxin production.

INTRODUCTION

Mycotoxins are secondary metabolites produced by fungi that occur naturally in foodstuffs. These toxins cause a large variety of toxic effects on vertebrates, including men [1]. Exposure occurs predominantly by the ingestion of contaminated food, mainly cereals used in the preparation of animal feed such as corn, wheat, peanuts and sorghum, as well as other raw materials [2].

Contamination of foodstuffs with toxigenic fungi may occur in different phases of production and processing, from sowing to transportation and storage. Besides, as mycotoxins are chemically stable, they may remain in the foodstuff even after fungi were removed by usual processing and packaging processes.

Diseases caused by mycotoxins are called mycotoxicosis, and are characterized by diffuse syndromes. However, depending on the type of toxin, lesions predominantly affect a given organ or tissue, such as liver, kidneys, epithelial tissue and Central Nervous System. The simultaneous occurrence of two or more mycotoxins is not rare, and toxic effects on the susceptible organism may be synergistic [3].

The economic impact of mycotoxins includes mortality of the animal, increased costs with veterinary care, and decreased production efficiency. In the poultry production acute mycotoxicosis is rare, and chronic exposure to low levels of mycotoxins is responsible for reduced productivity and increased susceptibility to infectious diseases [3].

In Brazil, several mycotoxins have been detected in food and feed, among them aflatoxins, fumonisins and ochratoxins are of great importance, not only because of their frequency of occurrence but also due to their toxigenic potential to human and animal [4].

Prevention of growth and mycotoxin production in fungi from the field is usually considered the best approach to impede the harmful effects of mycotoxins on animal and human health. So for lowering pre-harvest contamination, the treatment of field crops with fungicides is the traditional technique. The effect of fungicides on mold growth and mycotoxin biosynthesis is affected by several factors, including their chemical nature, rate of application, crop type, fungal species, and storage conditions [5].

Mycotoxin production is dependent on a number of factors, e.g., water activity of the stored product, temperature, presence of chemical preservatives, and microbial interactions. Water availability or moisture content is one of the most important factors in the prevention of fungal growth and mycotoxin production [6].

Although the prevention of mycotoxin contamination in the field is the main goal of the agricultural and food industries, the contamination of various commodities with *Aspergillus* or *Penicillium* isolates and their mycotoxins is unavoidable under certain environmental conditions. Postharvest strategies aim at lowering fungal contamination and consequently, the mycotoxin content of agricultural products during storage, handling, processing, and transport [7]. Such strategies include the improvement of drying and storage conditions, the use of chemical and natural agents, and irradiation. Several fungicides have been shown to be active in reducing either fungal growth or ochratoxin A (OTA) levels in grapevine, coffee and grain [8, 9]. However, it can induce resistance of the fungal against some fungicides, and also to the high toxicity of the substance is a problem to health and environmental.

Some natural compound and essential oils extracted from plants are capable of preventing OTA contamination in cereal [10]. Flavonoids, as rutin and quercetin, and antioxidant as caffeic acid were related as inhibitor of both growth and OTA synthesis by *A. carbonarius* [11, 12].

This chapter will focus on the use of plants extracts, essential oils and substances isolated from them to control ochratoxigenic fungi and ochratoxin production.

OCHRATOXINS

Ochratoxins were first isolated from the species *Aspergillus ochraceus* (previously known as *A. alutaceus*), and these toxins may be produced by several species in the genera *Aspergillus* and *Penicillium* [13]. The ochratoxin group has 7 components, but only OTA has been found worldwide as a natural contaminant of grains. The kidneys are the target organs of OTA, which interferes in the synthesis of macromolecules in cells of the renal parenchyma, including DNA, RNA and proteins. Besides, it affects renal carbohydrate metabolism, damaging the epithelium of the proximal tubules and causing osmotic diuresis by means of decreased absorption of electrolytes and increased water excretion [13].

Most frequently, OTA contaminates grains and cereals such as corn, barley, wheat and rye. Ochratoxin production occurs at room temperature between 4-37°C, relative humidity between 18.5 – 40.4% and water activity between 0.83 – 0.90, depending on the toxigenic lineage [7].

There are few surveys on the occurrence of OTA in foodstuffs and animal feed in Brazil. It should be pointed out, however, that given the size of the country, different regions have different climates. Southeastern Brazil, for instance, has optimum conditions for the production of OTA by *A. ochraceus* lineages, whereas the southern region has ideal conditions for the genus *Penicillium* [4].

Ochratoxins are among the most economically important of mycotoxins. The ochratoxin-producing fungi may contaminate agricultural products in the field, during storage, or during processing. Since contamination of food and feeds with ochratoxin can occurs despite the most strenuous efforts at prevention and since contamination can occur in the field before harvest, the development of the methods that degrade and prevent the mycotoxins formation are very important.

Substances and Extract or Essential Oil from Plants

Although the prevention of mycotoxins contamination in the field is the main goal of the agricultural and food industries, the contamination of various commodities with *Aspergillus* or *Penicillium* isolates and their mycotoxins are unavoidable under certain environmental conditions. Postharvest strategies aim at lowering fungal contamination and consequently, the mycotoxins content of agricultural products during storage, handling, processing, and transport. Such strategies include the improvement of drying and storage conditions, the use of chemical and natural agents, and irradiation [7].

Synthetic chemicals such as fungicides have been used for a long time and have greatly contributed in management of economic losses. The application of such chemicals has led to a number of environmental and health problems due to their residual toxicity, carcinogenicity, hormonal imbalance and spermatotoxicity [14, 15].

Because of indiscriminate use, some microorganisms have developed resistance to most widely used synthetic fungicides. Hence, there is a need to develop new fungicides with improved performance as well as ecofriendly in nature [16]. Some secondary metabolites of plants with complex molecular structures have antimicrobial properties. Studies have showed that secondary metabolites include alkaloids, flavonoids, isoflavonoids, biflavonoids, anthocianidins, β-glucans, tannins, coumarins, glycosides, terpenes and phenolic compounds are biologically active against mycotoxigenic fungal [7].

During recent years, substantial interest has developed in the preservation of grains by the use of essential oils to effectively retard mycotoxins production as well as the fungal growth. Different extracts and essential oils of spices, herbs and other plant materials are becoming increasingly important in the food industry because of their antifungal properties.

The increased demand for safe and organic food, without chemical preservatives, provokes many researchers to investigate the antimicrobial effects of natural compounds. Numerous investigations have confirmed the antimicrobial action of essential oils in model food systems and in real food. There is evidence that essential oils are more strongly antimicrobial than is accounted for by the additive effect of their major antimicrobial components; minor components appear, therefore, to play a significant role. Essential oils are considered safe, the possibility of reinforcing their natural antimicrobial effects by the addition of small amounts of other natural preservatives may be a way to attain a balance between sensory acceptability and antimicrobial efficacy [17].

Natural substance isolated from extract and essential oil has been tested for antifungal activity against *P. verrucosum, A. ochraceus, A. carbonarius, A. niger*. Substances as thymol, carvacrol, menthol, resveratrol, rutin, quercetin and caffeic acid, cinnamaldehyde, cinnamon, citral, eugenol and anisaldehyde inhibited *P. verrucosum* and *A. ochraceus* growth and on ochratoxin A production [12, 18, 19, 20, 21].

Essential oils extracted from plants, such as *Thymus vulgaris, Aframomumdanielli* [22, 23], cinnamon, and clove leaf [24], which affect mold growth and OTA synthesis. Other essential oils extracted from plants (thyme, cinnamon, marigold, spearmint, basil, quyssum, caraway, and anise) [25], and those of oregano, mint, basil, sage, and coriander, have also been found to inhibit the growth of ochratoxigenic fungi, while oregano and mint oils also inhibited OTA production by *A. westerdijkiae* [18]. The efficacy of three essential oils (bay, clove and cinnamon) and the antioxidant resveratrol (0–500 μg/g) on the control of growth and ochratoxin A (OTA) production by *P. verrucosum* and *A. westerdijkiae* under different water activity (0.90, 0.95 and 0.995), and temperature (15 and 25°C) conditions on irradiated wheat grain was determined by Aldred et al. [18].

Romero et al. [12] described the inhibition of growth and OTA biosynthesis by *A. carbonarius* by flavonoid and nonflavonoid compounds. Alkaloids produced by *Piper longum*, and components of sesame oil and turmeric have also been found to suppress both fungal growth and ochratoxin production [7, 26]. The essential oils of plant species mainly of the species of family Labiateae, Compositeae, Lamiaceae, Apiaceae, Asteraceae, Lauraceae, Piperaceae were related as antifungal on ochratoxigenic strains (Table 1).

Essential oils of *Aframomumdanielli* were able to reduce OTA to zero level at 2000 ppm while the aqueous extracts were able to reduce OTA to less than 20 ppb. The reduction efficiency of both the extracts and the essential oils were within 64.28 - 95% while that of the aqueous extract was 5.6-80% [22].

The essential oils of species of *Citrus* inhibited on the growth of moulds commonly associated with food spoilage [27]. The work showed the capacity to reduce or inhibit the growth of the molds *P. chrysogenum*, *P. verrucosum*, *A. niger* and *A. flavus* by *Citrus lemon, C. reticulate, C. paradis and C. sinensis* during the nine days. The *A. niger* and *A. flavus* growth were completely inhibited at concentration 0.94% of any of the essential oils used.

The essential oil from *Ocimumgratissimum*, *Thymus vulgaris* and *Cymbopogoncitratus* were the most effective and prevented conidial germination and the growth of *P. expansum*, *P. verrucosum* and *A. ochraceus*, respectively [28].*C.citratus* essential oil at 0.33 µL/mL caused 100% mycelial inhibition against the fungi *A. niger*, and *A. ochraceus*. Oil was fungistaticat 0.29 µL/mL.*C. citratus* oil completely inhibited the growth and mycotoxins production (aflatoxin B_1 and OTA) [29].

Artemisia nilagirica essential oil exhibited a broad range of fungitoxicity against other grape berry-rotting fungi of the species of *Aspergillus*. *A. nilagirica*oil completely suppressed the fungal growth and the mycotoxins production (aflatoxin B_1 and OTA) [30].

Essential oils from oregano (*Origanumvulgare* L.), were tested for antifungal activity against *P. verrucosum*. The lowest minimal inhibitory concentrations (MICs) were achieved for essential oil of oregano. During 21 days of exposure, the growth of *P. verrucosum*, and subsequently the production of OTA, was fully inhibited. The observed antifungal effects depended on the agent, the concentration used, and the time of interaction between the agent and *P. verrucosum* [31].

Table 1. Extract and essential oil from plant species with antifungal activity and inhibition of ochratoxin production

Plants	Extract/oil	Molds	Toxin	Reference
Adenocalymmaalliaceum	Extract	*A. niger*		34
Aframomumdanielli	Essential oil and aqueous extract		OTA	22
Aloysiatriphylla		*A. alutaceus*		35
Artemisia nilagirica	Essential oil	*A. ochraceus*		29
Baccharisarticulata		*A. alutaceus*		35
Baldeaboldus		*A. alutaceus*		35
Cecropiaadenopus		*A. alutaceus*		35
Chenopodiumambrosioides	Essential oil	*A. niger*		15
Cinnamomumzeylanicum	Essential oil	*A. niger*		36
Citrus lemon	Essential oil	*A. niger* *A. flavus* *P. verrucosum* *P. chrysogenum*		27
C. paradisi	Essential oil	*A. niger* *A. flavus* *P. verrucosum* *P. chrysogenum*		27
C. reticulata	Essential oil	*A. niger* *A. flavus* *P. verrucosum* *P. chrysogenum*		27
C. sinensis	Essential oil	*A. niger* *P. verrucosum*		27
Cymbopogoncitratus	Essential oil Extract	*A. alutaceus* *P. verrucosum* *A. niger*		30,37,38,3 9,40,41,42, 43
Equisetum giganteum		*A. alutaceus*		35
Foenicculumvulgare	Essential oil	*A. alutaceus*	OTA	44
Mentaviridicum	Essential oil	*A. alutaceus*	OTA	44
Lippiaturbinata	Essential oil	*A. niger*		32
Ocimumbasilicum	Essential oil	*A. alutaceus*	OTA	44
O. gratissimum	Essential oil	*A. alutaceus* *P. verrucosum*		39,40
Oreganumvulgare	Essential oil	*P. verrucosum*	OTA	31
Passifloracoerulea		*A. alutaceus*	OTA	35
Peumusboldus	Essential oil	*A. alutaceus*	OTA	44
Pimpinellaanisum	Essential oil	*A. alutaceus*	OTA	44
Syzygiumaromaticum	Essential oil	*A. alutaceus*	OTA	44
Thymus eriocalyx	Essential oil	*A. niger*		45,46
T. serpyllum	Essential oil	*A. ochraceus* *A. niger* *A. carbonarius*		33
T. vulgaris	Essential oil	*P. verrucosum* *A. niger* *A. alutaceus*	OTA	39,44,46,4 7
T. x-porlock	Essential oil	*A. niger*		46,47

Essential oils from boldo (*Pëumusboldus* Mol.), poleo (*Lippiaturbinata* var. integrifolia), clove (*Syzygiumaromaticum* L.), anise (*Pimpinellaanisum*) and thyme (*Thymus vulgaris*) were evaluated for their effectiveness against the growth of *A. niger* aggregate and accumulation of OTA. The evaluation was performed by compound dissolution at the concentrations of 0, 500, 1500 and 2500 µL/L in peanut meal extract agar (PMEA) and exposure to volatiles of boldo, poleo (0, 1000, 2000 and 3000 µL/L) and clove oils (0, 1000, 3000 and 5000 µL/L), taking into account the levels of the water activity (aw) of the medium (aw 0.98, 0.95, 0.93). The antifungal efficiency of the oils tested was observed in vapor exposure assay. At all aw levels, complete fungal growth inhibition were achieved with boldo at concentrations of 1500 and 2000 µL/L by contact and volatile assays, respectively. OTA accumulation was closely dependent on aw conditions [32].

Thymus serpyllum L. and of its components thymol and total phenols (TPC) extracted from the plant were evaluated on the growth and mycotoxin production of *A. ochraceus*, *A. carbonarius*, and *A. niger*. Minimal inhibitory concentration (MIC) determined for the essential oil and thymol, and selected concentration of the TPC extract inhibited fungal growth and OTA biosynthesis by more than 60%, depending on the conditions and duration of incubation with the fungi [33].

Conclusion

Ochratoxin contamination of food and feed poses a serious health hazard to animals and humans. Due to OTA toxicity, and in order to assure human and animal health, this toxin is not allowed to be present above maximum permitted levels in agricultural products intended to be used as foods or animal feed.

The choice of the any method to mycotoxins control must considered the cost, facility of acquisition and implementation.

The procedures that prevent the formation of mycotoxins or fungal contamination represent a good way to avoid the exposure to animal feed and foods for human consumption. The use of natural compounds, like essential oils, with lower toxicity of the synthetic products can be a good alternative to the mycotoxins control.

However, it is important remember that the best way of mycotoxin control any is the prevention using the HACCP (Hazard Analysis and Critical Control

Points) studies in all segments of food and feed channel. Good Agricultural Practices (GAP), including Good Hygienic Practices (GHP) to make sure that the harvested grains will not present a hazard to the animal and consumer health and also minimize the risk to exposure to any kind of mycotoxins.

REFERENCES

[1] Coulombe, RA. Aflatoxins. In: Sharma, RP; Salunkhe, DK (editors). *Mycotoxins and phytoalexins*. Boca Raton: CRC Press, 1991, pp. 103-143.

[2] Chu, FS. Mycotoxins: food contamination, mechanism, carcinogenic potencial and preventive measures. *Mutation Research*, 1991, 259, 291-306.

[3] Hussein, H S; Brasel, JM. Toxicity, metabolism, and impact of mycotoxins on humans and animals. *Toxicology*, 2001, 167,101–134.

[4] Oliveira, CAF; Gonçalez, E; Reddy, KRN. Mycotoxicosis in domestic fowl. In: Gonçalez, E, Felicio, JD, Aquino, S (editors). *Mycotoxicoses in animals economically important*. New York: Nova Science Publishers, Inc., 2012, pp 97-115.

[5] Kabak, B; Dobson, ADW; Var, I. Strategies to prevent mycotoxin contamination of food and animal feed: a review. *Critical Reviews in Food Science and Nutrition*, 2006, 46, 593–619.

[6] Abramson, D. Mycotoxin formation and environmental factors. In Sinha, KK, Bhatnagar, D (editors). *Mycotoxins in Agriculture and Food Safety*. New York: Marcel Dekker, 1998; pp. 255–277.

[7] Varga, J; Kocsubé, S; Péteri, Z; Vágvölgyi, C; Tóth B. Chemical, Physical and Biological Approaches to Prevent Ochratoxin Induced Toxicoses in Humans and Animals. *Toxins*, 2010, 2, 1718-1750.

[8] Bellí, N; Marní, S; Sanchis, V; Ramos, AJ. Impact of fungicides on Aspergillus carbonarius growth and ochratoxin A production on synthetic grape-like medium and on grapes. *Food Additives and Contaminants*, 2006, 23, 1021–1029.

[9] Visconti, A; Perrone, G; Cozzi, G; Solfrizzo, M. Managing ochratoxin A risk in the grape-wine food chain. Food Additives and Contaminants, 2008, 25, 193–202.

[10] Felicio, JD; Aquino, S; Goncalez, E. Methods for preventing and controlling mycotoxins and mycotoxigenic fungi. In: Gonçalez, E,

Felicio, JD, Aquino, S (editors). *Mycotoxicoses in animals economically important*. New York: Nova Science Publishers, Inc., 2012, pp 143-183.

[11] Basilico, MZ; Basılico, JC. Inhibitory effects of some spice essential oils on *Aspergillusochraceus* NRRL3174 growth and ochratoxin A production. *Letters Applied in Microbiology*, 1999, 29, 238–241.

[12] Romero, SM; Alberto, MR; de Nadra MCM; Vaamonde, G. Inhibition of growth and ochratoxin A biosynthesis in Aspergillus carbonarius by flavonoid and nonflavonoid compounds. *Mycotoxin Research*, 2009, 25, 165–170.

[13] Leeson, S; Diaz, GJ; Summers, JD. *Poultry metabolic disorders and mycotoxins*. Guelph: University Books, 1995, p. 352.

[14] Pandey, A. Solid state fermentation. *Biochemical Engineering Journal*, 2003, 13, 81-84.

[15] Kumar, R; Mishra, AK; Dubey, NK; Tripathi, YB. Evaluation of *Chenopo-diumambrosioides* oil as a potential source of antifungal, antiaflatoxigenic and antioxidant activity. *International Journal of Food Microbiology*, 2007, 115, 159-164.

[16] McKenzie, KS; Sarr, AB; Mayura, K; Bailey, RH; Miller, DR; Rogers, TD; Norred, WP; Voss, KA; Plattner, RD; Kubena, LF; Phillips, TD. Oxidative degradation and detoxification of mycotoxins using a novel source of ozone. *Food Chemistry and Toxicology*, 1997, 35, 807–820.

[17] Rasooli I; Fakoor MH; Yadegarinia D; Gachkar L; Allameh A; Rezaei MB. Antimycotoxigenic characteristics of *Rosmarinusofficinalis* and *Trachyspermu-mcopticum* L. essential oils, *International Journal of Food Microbiology*, 2008, 122, 135-139.

[18] Alched, D; Cairns-Fuller, V; Magan, N. Environmental factores affect afficaacy of some essential oils and resveratrol to control growth and ochratoxin A production by Penicilliumverrucosium and Aspergilluswesterdijkiae on wheat grain. *Journal of Stored Products Research,* 2008, 44, 341-346.

[19] Hua, H; Xing, F; Selvaraj, JN; Wang, Y; Zhao, Y; Zhou, L; Liu, X; Liu, Y. Inhibitory Effect of Essential Oils on *Aspergillusochraceus* Growth and Ochratoxin A Production. e108285, 2014, doi:10.1371/ journal.pone.0108285

[20] Vazquez, BI; Fente, C; Franco, CM; Vazquez, MJ; Cepeda, A. Inhibitory effects of eugenol and thymol on *Penicilliumcitrinum*strains in culture media and cheese. *International Journal of Food Microbiology*, 2001, 67, 157-63.

[21] Palumbo, J; O'Keeffe, T; Mahoney, N. Inhibition of ochratoxinA production and growth of Aspergillus species by phenolic antioxidant compounds. *Mycopathologia,*. 2007, 164, 241–248.

[22] Aroyeun, SO; Adegoke, GO. Reduction of ochratoxin A (OTA) in spiked cocoa powder and beverage using aqueous extracts and essential oils of *Aframomumdanielli. African Journal Biotechnology*, 2007, 6, 612–616.

[23] Nguefack, J; Nguikwie, SK; Fotio, D; Dongmo, B; Zollo, PHA; Leth, V. Fungicidal potential of essential oils and fractions from Cymbopogoncitratus, Ocimumgratissimum and Thymus vulgaris to control Alternariapadwickii and Bipolarisoryzae, two seed-borne fungi of rice (Oryza Sativa L.). *Journal of Essential Oil Research*, 2007, 19, 581–587.

[24] Magan, N; Aldred, D. Postharvest control strategies: minimizing mycotoxins in the food chain. *International. Journal of Food Microbiology*, 2007, 119, 131–139.

[25] Solimar, KM; Badeaa, RI. Effect of oil extracted from some medicinal plants on different mycotoxigenic fungi. *Food and Chemical Toxicology*, 2002, 40, 1669–1675.

[26] Lee, SE; Park, BS; Bayman, P; Baker, JL; Choi, WS; Campbell, BC. Suppression of ochratoxin biosynthesis by naturally occurring alkaloids. *Food Additives and Contaminants Part A*, 2007, 24, 391–397.

[27] Viuda-Martos, M; Ruiz-Navajas, Y; Fernández-López, J;Perez-Álvarez, J. Antibacterial activity of lemon (*Citrus lemon* l.), mandarin (*Citrus reticulata* l.), grapefruit (*Citrus paradisi* l.) and orange (*Citrus sinensis* l.) essential oils. *Journal of Food Safety*, 2009, 28, 567–576.

[28] Nguefack, J; Dongmo, JBL; Dakole, CD; Leth, V; Vismer, HF; Torp, J; Guemdjom, EFN; Mbeffo, M; Tamgue, O; Fotio, D; Zollo, PHA; Nkengfack, AE. Food preservative potential of essential oils and fractions from Cymbopogoncitratus, Ocimumgratissimum and Thymus vulgaris against mycotoxigenic fungi. *International Journal of Food Microbiology*, 2009, 131, 151-156.

[29] Sonker, N; Pandey. AK; Singh, P. Efficiency of *Artemisia nilagirica*(Clarke) Pamp. essential oil as a mycotoxicant against postharvest mycobiota of table grapes.*Journal of the Science and of Food Agriculture*,2015, 95, 1932-1939.

[30] Sonker, N; Pandey, AK; Singh, P; Tripathi, NN. Assessment of *Cymbopogon-citratus*(DC.) Stapf Essential Oil as Herbal Preservatives Based on Antifungal, Antiaflatoxin, and Antiochratoxin Activities and *In*

Vivo Efficacy during Storage. *Journal of Food Science*, 2014, 79, 620-634.

[31] Jersek, B; Ulrih, NP; Skrt, M; Gavarić, N; Bozin, B; Mozina, SS. Effects of selected essential oils on the growth and production of ochratoxin A by *Penicilliumverrucosum*. *ArhivzaHigijenuRadaToksikologiu*, 2014, 65, 199-208.

[32] Passone, M; Girardi, NS; Etcheverry, M. Evaluation of the control ability of five essential oils against *Aspergillus* section Nigri growth and ochratoxin A accumulation in peanut meal extract agar conditioned at different water activities levels. *International Journal of Food Microbiology*, 2012, 159, 198–206.

[33] Sokolic-Mihalak, D; Frece, J; Slavica, A; Delas, A; Pavlovic, H; Markov, K. The effects of wild Thyme (*Thymus serpyllum*l.) essential oil components against ochratoxin-producing *Aspergilli*.*ArhivzaHigijenuRadaToksikologiu*, 2012, 63, 457-462.

[34] Shukla, R; Kumar, A; Prasad, CS; Srivastava, S; Dubey, NK. Antimycotic and antiaflatoxigenic potency of Adenocalymmaalliaceum-Miers.on fungi causing biodeterioration of food commodities and raw herbal drugs. *International Biodeterioration& Biodegradation*, 2008, 62, 348-351.

[35] Rizzo, I; Vedoya, G; Maurutto, S; Haidukowski, M; Varsavsky, E. Assessment of toxigenic fungi on Argentinean medicinal herbs. *Microbiological Research*, 2004, 159, 113–120.

[36] Singh, HNP; Prasad, MM; Sinha, KK. Efficacy of leaf extracts of some medicinal plantas against disease deve lopment in banana. *Letters in Applied Microbiology,* 1993, 17, 269-271.

[37] Mishra, AK; Dubey, NK. Evaluation of some essential oils for their toxicity against fungi causing deterioration of stored food commodities. *Applied and Environmental Microbiology*, 1994, 60, 1101-1105.

[38] Nguefack, J; Leth, V; Zollo, PHA; Mathur, SB. Evaluation of five essential oils from aromatic plants of Cameroon for controlling food spoilage and mycotoxin producing fungi. *International journal of food microbiology*, 2004, 94, 329-334.

[39] Nguefack, J; Dongmo, JBL; Dakole, CD; Leth, V; Vismer, HF; Torp, J; Guemdjom, EFN; Mbeffo, M; Tamgue, O; Fotio, D; Zollo, PHA; Nkengfack, AE. Food preservative potential of essential oils and fractions from *Cymbopogoncitratus*, *Ocimumgratissimum* and *Thymus vulgaris* against mycotoxigenic fungi. *International Journal of Food Microbiology*,2009, 131, 151-156.

[40] Bluma, RV; Etcheverry, MG. Application of essential oils in maize grain: Impact on *Aspergillus* section *Flavi* growth parameters and aflatoxin accumulation. *Food Microbiology*, 2008, 25, 324-334.

[41] Kumar A; Shukla R; Singh P; Prasad CS; Dubey NK. Biodeterioration of some herbal raw materials by storage fungi and aflatoxin and assessment of *Cymbopogonflexuosus*essential oil and its components as antifungal. *International, Biodeterioration& Biodegradation*,2009, 63, 712-716.

[42] Soliman, KM; Badeaa, RI. Effect of oil extracted from some medicinal plants on different mycotoxigenic fungi. *Food and Chemical Toxicology*, 2002, 40, 1669-1675.

[43] Bluma, RV; Amaiden, MR; Daghero, J;Etcheverry, MG.Control of *Aspergillus* section *Flavi* growth and aflatoxin accumulation by plant essential oils. *Journal of Applied Microbiology*, 2008, 105, 203–214.

[44] Rasooli, I; Rezaei, MB; Allameh, A. Growth inhibition and morphological alterations of *Aspergillusniger* by essential oils from *Thymus eriocalyx* and *Thymus x-porlock*. *Food Control*, 2006, 17, 359–364.

[45] Kumar, A; Shukla, R; Singh, P; Prasad, CS; Dubey, NK. Assessment of *Thymus vulgaris* L. essential oil as a safe botanical preservative against post-harvest fungal infestation of food commodities. *Innovative Food Science & Emerging Technologies*, 2008, 9, 575-580.

INDEX

A

D

E

F

I

J

K

L

M

N

O

P

Q

R

S

T

U

V

W

Y